Dark Matter

The Displacement of an Object in a Field Density

(Dark Matter Paths to Continuous Fusion Through Coalescent Physics)

Carolyn Edgar

Published By **Bengion Cosalas**

Carolyn Edgar

All Rights Reserved

Dark Matter: The Displacement of an Object in a Field Density (Dark Matter Paths to Continuous Fusion Through Coalescent Physics)

ISBN   978-1-77485-876-9

No part of this guidebook shall be reproduced in any form without permission in writing from the publisher except in the case of brief quotations embodied in critical articles or reviews.

Legal & Disclaimer

The information contained in this ebook is not designed to replace or take the place of any form of medicine or professional medical advice. The information in this ebook has been provided for educational & entertainment purposes only.

The information contained in this book has been compiled from sources deemed reliable, and it is accurate to the best of the Author's knowledge; however, the Author cannot guarantee its accuracy and validity and cannot be held liable for any errors or omissions. Changes are periodically made to this book. You must consult your doctor or get professional medical advice before using any of the suggested remedies, techniques, or information in this book.

Upon using the information contained in this book, you agree to hold harmless the Author from and against any damages, costs, and expenses, including any legal fees potentially resulting from the application of any of the information provided by this guide. This disclaimer applies to any damages or injury caused by the use and application, whether directly or indirectly, of any advice or information presented, whether for breach of contract, tort, negligence, personal injury, criminal intent, or under any other cause of action.

You agree to accept all risks of using the information presented inside this book. You need to consult a professional medical practitioner in order to ensure you are both able and healthy enough to participate in this program.

TABLE OF CONTENTS

Chapter 1: Scale Of The Universe 1

Chapter 2: The Big Bang Model 16

Chapter 3: Inflation 37

Chapter 4: Dark Matter 58

Chapter 5: Dark Energy 83

Chapter 6: Dark Gravity............................... 98

Chapter 7: What Exactly Is An Hologram? ..116

Chapter 8: The Ether: A Story Of A Old Theory 127

Chapter 9: Future Of The Universe.............154

Conclusion 181

Chapter 1: Scale Of The Universe

Modern Science

Homo sapiens has existed for thousands of years. And through all of that time up to the present century humanity has not been able to gain a complete understanding of the universe. This is due to the advancement of technology that allows for massive ground-based telescopes, sensitive electronic detectors, super-fast computers to analyze the information gathered and also the capability to launch scientific satellites for astronomical research into space that we've been able to discover the mysteries of cosmology as well as our understanding of the Big Bang origin of the universe.

The study of Cosmology encompasses the characteristics of the universe in general and the understanding that the Universe changed over the course that spans thousands of millions of years. The evolution of the physical laws have been accompanied by modern observational techniques that permit us to continually increase and improve our understanding of the universe.

In the early days of Babylonian Cosmology, the universe was believed to be centered around the god of gods and believed to exist multiple earths and many heavens. In the ancient Hindu Cosmology, a golden egg was believed to be the source of our universe. there are cycles that last 311 trillion years, which have periods of expansion and collapse that are followed by the rebirth. In this cosmology at the same time there are numerous universes, maybe many.

The Greeks were apprehensive about the various theories. Philolaus was, prior to Aristotle was among the first people in the West to propose an non-geocentric (non-Earth-centric) conception of the world and a view of the Earth revolved around the "central flame". Aristarchus was a philosopher who lived following Aristotle recognized his Sun as the primary fire in a Heliocentric Cosmology. In his cosmology, the fixed stars were placed around the Sun together with the planets.

The model was proposed by Aristotle during the fourth century BC who suggested an spherical earth as the central point of the static eternal Cosmology. The Earth in the model was enclosed by celestial spheres that were each focused on the Earth's location. Ptolemy followed Aristotle by several centuries, enlarged the model using intricate epicycles (circles that were attached to circles) to explain the movements of planets. The Aristotelian universe had infinite temporal dimension. Christian, Jewish and Muslim philosophers of the Middle Ages developed arguments for an infinite temporal dimensions with a clear beginning. Different versions that are based on the Ptolemaic model can be seen throughout Persian as well as Indian writings from the Middle Ages.

Unfortunately, the Aristotelian-Ptolemaic view held sway for a 2000- year period, slowing scientific progress. This modern era of science and the rapid advancement of knowledge started around 500 years ago. It is usually referred to as beginning with Copernicus. In the final days of his life, in 1543 Nicolaus Copernicus released his Heliocentric (Sun-centered) theory of the solar system. Tycho Brahe was a co-author of Copernicus's work and, after his death, suggested a hybrid model that included the planets orbiting around the Sun as well as the Sun in orbit around the Earth. In actuality the

Copernican Heliocentric model wasn't popular until about the time of the year 1770. Of course, the Copernican model, which was refined to be codified in physical law in the works of Kepler, Newton and others didn't address the location that the Sun in the galaxy or whether we reside in a single galaxy , or are part of one of the billions of galaxies that exist in the universe.

In the past 2000 years, our scientific understanding has improved using improved observational techniques and a greater understanding about the scientific laws have been able to dispel one myth after the other. We used to believe we believed that Earth was the centre of the universe in addition to the Sun was distinct than other star systems. In the last 100 years, we reversibly changed that perception and began mapping our galaxy. Initially, the idea was to be part of the universe. It was only in the last century did it become clear we were not alone. Milky Way galaxy is one out of billions. It was thought that the universe was static, but we have now discovered that the universe as a whole is expanding quickly. Galaxies generally fly between each other with speeds that vary between a few kilometers all the way to close to what the velocity of light. In the majority in the twentieth century,, we believed that there had been a Big Bang explosion and that the expansion of the universe was slowing due to gravity. Today we know that the universe's expansion is currently increasing in speed!

Understanding what is the Scale of the Universe

The table below gives distances to various objects in the astronomical universe, measured in parsecs, kms and length of the light, which represents the length of time that light travels passing through the objects to Earth. A parsec, which is described in the next chapter, is a common measurement used by astronomers. It is roughly 3.26 light years. The dynamic spectrum of distances is vast that spans from a single second in travel the Moon up to 14 billion years prior to the beginning of our universe It is an astronomical range that is around 400 million billion billion.

Table 1.1 Distances to celestial objects. The globular cluster can be described as a tight-knit collection of stars in the framework of a galaxy. The cluster of galaxies are a gravitationally bound collection comprising hundreds or thousands galaxies. Our universe, also known as at least the portion which we can see is truly vast. The galaxies with the most distance are three billion times farther away than the closest star.

A realization of the magnitude of our universe is not too recent, in the last 100 years, essentially. Before the period of Copernicus it was considered that Earth was the centre of the universe. Also, stars were believed to be farther away than they actually were. In 1514, he began to develop the heliocentric model, and in 1543 published the Heliocentric Model, which provided an accurate description of the solar system, which included that of the Sun, Earth, Moon and the planets that were then identified. The event is believed to be the start of the Scientific Revolution of the past more than 500 years. The Copernican Heliocentric model came to the rescue of the geocentric view of Ptolemy that has been in place for nearly two millennia. The geocentric model was such a solid belief it Galileo found himself forced to change his stance in favor of the heliocentric perspective and placed under house detention under his Catholic Church throughout the rest period of his existence. In a bizarre twist, Copernicus was the one who developed his model also a priest.

Although the Greeks tried to determine the distance between the Sun and Earth using various methods but the first reliable measurements occurred during the 1660s. Through measuring the movement from Venus over the Solar disk at two different locations, one could utilize geometry to determine the distance of the Sun in relation to Earth. The most accurate estimate of the time period, made by Christiaan Huygens was within 2.5 percent of the actual value.

We can now measure the distances of our personal Solar System extremely precisely by using radar to determine for instance, the distance between Earth and Venus at various dates, as well as using orbital calculations. This technique is however exclusively within our Solar System.

The measurement of distances to objects in astronomy outside the Solar System has been one of the most difficult problems in Astronomy. We can gauge how bright an object appears however, we do not have an method to determine the absolute brightness of an object that we refer to as luminosity. The absolute brightness, or luminosity of stars and other objects can be extremely different. If we can determine both the apparent brightness as well as the luminosity, we will calculate the distance, as proportion of apparent luminosity and brightness is dependent on the magnitude of the distance between the object.

Parallax

The most straightforward method to calculate the distance from a star is to use the parallax method. This method is only applicable for the closest stars. The method is to determine the apparent direction of the star regards to background stars at different dates of the year. the most ideal scenario is six months apart. This will give a reference point that is equal to the size of planet's orbit in relation to the Sun and the desired star will appear to shift little bit in comparison to those stars.

The distance between the two locations on the Earth is two Astronomical units (AU). This is the apparent direction of the nearest star (purple dot close to middle) is different from the background stars based on the time of the year in which the Earth's location changes.

The term "parallax" refers to the distance between Earth and Sun that is approximately 150 million km. This distance is also known as 1 AU, which is an the astronomical unit. A arcsecond is 1/3600 of an angle degree. If an object's position changes by one arcsecond, for an measurement baseline of 1 AU, the distance is measured in parsecs or approximately 3.26 Light years. Light years are the length that light travels over one year. The closest celestial body, Proxima Centauri, is 1.3 parsecs or 4.24 light years away.

As we progress to more far-off stars of our galaxy and later to distant extragalactic distances the methods used to measure distances become more diverse, and in general, less precise. The collection of methods is called cosmic distance ladder. We begin by measuring objects closer to us and then employ different methods to make them jump further in distance. To allow a variety of these techniques to be effective, we need to locate something with an intrinsic brightness or luminosity. This is a kind of objects the standard candle. Once we know the exact brightness of the standard candle we can calculate its distance by taking the apparent brightness.

Our Milky Way Galaxy

It's been a challenging endeavor in mapping our universe also known as The Milky Way, in part because we live within it. We don't have the benefit of seeing our Milky Way from the outside as a large amount of dust and gas obscures our vision towards the middle in the galaxies. The Greeks believed that they believed that the Milky Way might consist of distant stars that are not resolved. In around 1000 AD, Alhazen in Basra and Cairo tried to determine the distance from the Milky Way center with geometric methods and discovered that it was very far away.

Galileo became the very first person to find Galileo was the first to see Milky Way galaxy by telescope and to observe that it is made up of numerous tiny stars. Immanuel Kant during the 18th century, proposed that the Milky Way is a rotating body made up of a numerous stars, that are held together by gravity. In this theory, the scientist was correct. Later , in the same century William

Herschel began to sketch and map his idea of the Milky Way.

We have discovered that our galaxy is home to more than 100 billion stars and it is estimated that our Sun is only 25,000 light years of the central region. Evidence suggests that the center has an enormous black hole that is just a few millions of solar masses.

Our Galaxy isn't alone.

The most significant breakthrough in the understanding of the size of the universe came with construction of the world's biggest telescope, in 1919, located on Mt. Wilson located in California. The size of the telescope is determined by the dimension of the mirror (or lens) and the Hooker telescope measured 100 inches in diameter, which is 2.5 meters. Prior to the construction installation of the telescope majority view among Astronomers was that our galaxy was the universe itself, and that the various Nebulae (fuzzy areas that emit light) were dust and gas clouds in our galaxy. Although some nebulae do exist found within of the Milky Way, a huge majority are now believed to be extraterrestrial They are galaxies identical to our own and comprise billions of stars, and dust and gas.

There are over 10,000 galaxies. Many are billions of light-years distant. The majority of the objects that appear in the image, including small little ones, are distant galaxies. They are not stars or objects from our galaxy. (Credit: NASA / STSci)

In the 1920s, Edwin Hubble worked at the Mt. Wilson Observatory to demonstrate that spiral nebulae existed farther away than other regions of our galaxy. He discovered within the Andromeda Nebulae (we are now referring to this as The Andromeda galaxy) an extremely specific kind of variable star that we could calibrate to determine the total luminosity. This was Hubble's normal candle, which allowed Hubble to discover how far Andromeda was far enough from the galaxy to belong in the Milky Way. It is believed that the Milky Way, the Andromeda galaxy and the Magellanic Clouds and a few nearby galaxies do belong to what's called The Local Group, and it appears that this Local Group is bound together by gravitational force.

It is the Age of the Universe

Scientists and radioactive dating techniques have concluded an age for the Earth is approximately 4.54 billion years. This puts a lower limit to the universe's age. The oldest meteorites known to exist found in our solar system ranges from around 4.57 billion years, which means that it is likely it is possible that it is likely that the Sun as well as Earth were created around the same time.

The universe is much older than both the Sun as well as Earth. We are aware that the Sun isn't a first generation star because it has a large amount of elements that are heavier that hydrogen and Helium, and none of the heavier elements came into existence prior to the Big Bang. The heavier elements were created by nuclear fusion processes within stars by themselves, meaning that the Sun is likely to have developed after other stars have passed through their own life cycles. How do we be sure that the existence of the Big Bang?

The general relativity theory formed the basis for the creation of the Big Bang model. General relativity was first proposed in 1915. It was confirmed by observations of the bending of stars because of the gravity of the

Sun during the solar eclipse in 1919. The most remarkable proof of the general relativity on the size of the universe in general was discovered in 1929.

Two years before, Georges Lemaitre had derived an expanding universe by analyzing the general relativity equations and pioneered his Big Bang theory, by presuming a universe of constant density and uniformity at any point in moment. It has proved to be an incredibly accurate theory. Lemaitre was educated at Harvard College Observatory and the University of Cambridge and Harvard College Observatory and was ordained a priest in his home country, Belgium. In 1927, he released "A homogeneous universe that has constant mass and a growing radius, which accounts for the increasing radiatal velocity of the extragalactic nebula," and afterward was given the honor of a PhD in The Massachusetts Institute of Technology. Einstein was at first opposed to Lemaitre's theories. Alexander Friedmann in Russia had also suggested a similar solution at the time of 1922 however, he passed away at an early years of age, which meant that he wasn't as well-known. Furthermore, Lemaitre went further with his theories to try in his research to understand the redshift in distant galaxies.

The evidence from observation was taken to prove Lemaitre and Friedmann to be correct. This leads to discuss Lemaitre's Big Bang model in the next chapter.

Chapter 2: The Big Bang Model

Introduction

Everything that is now our universe was created 13.8 billion years ago beginning from the size of an atomic nucleus in the event known by the name of Big Bang. The evolution the understanding we have of modern cosmology happened through the 20th century. this era of modernity began by introducing general relativity around 1915. In 1929, just 14 years after - we were able to observe that the universe was increasing in a consistent way as a naturally occurring result out of the Big Bang.

We can actually see the remnant radiation from the expansion - this was first discovered through Arno Penzias, and Robert Wilson in 1964[4]. The crucial inflation theory, that added an understanding to the Big Bang and resolved several unanswered questions, was developed by the late 1980s. The concept of "missing material" (also known as "dark matter" was first developed in the 1930s and onwards. At the end in the twentieth century,, we discovered a convincing evidence for a second extremely important aspect - the "dark energy" that adds acceleration to the expansion of the universe. In a bizarre way, Einstein was the one who predicted, but later denied the dark energy referred to by the name of Cosmological Constant as a possible element in his general relativity equations. His most famous mistake is now true, and it was not even a mistake.

Cosmological Redshift

Light travels at a constant velocity in a vacuum. It is distinguished in terms of frequency (or wavelength). Atoms absorb or emit light at specific frequencies, which are determined by quantum levels (particular orbitals, as they are known as) of electrons inside the atom. These specific frequencies act as a reference point when we study the spectra of galaxies in the vicinity of the boundaries of our Milky Way galaxy.

What is interesting is that galaxies outside our area are receding away out of their position in the Milky Way at a fairly speedy rate. Also, the speed increases with distance. The further from the Milky Way is it!

Vesto Slipher along with other astronomers first noticed the shift in frequency of light coming from galaxies that were not visible in the early 1910s. The cause is explained by special relativity and is based on the speed of the observed galaxy in relation to Earth. The magnitude of the shift will be proportional to V/c, which is the velocity of the galaxy's retreat of Earth across the line of vision. A given line of spectral radiation will shift to a different frequency or wavelength. The spectrum radiation also changes in frequency.

The frequency shift can be expressed by the equation

$dn/n = -V/c$

In this equation, dn represents the frequency shift, n is an frequency that is unshifted (for the case of zero velocity) V is the speed in the direction of sight, and c is that of the speed at which light travels. The minus sign is a result due to the fact that velocity is considered to be positive towards us, and frequency is reduced in the event that you are moving towards the stars. Therefore, if that recession speed is one percent that of the light speed then the light will shift by 1% towards lower frequencies.

Light shifts towards blue (blueshift) or higher frequencies in the case of a galaxy close to us. In the event that it is moving away from us, as is the normal situation is that light will shift towards the red and lower frequency. The movement of the galaxy in a direction that is parallel to the line of sight does not cause an increase in blueshift or redshift; only the movement along the line of sight is important. While some nearby galaxies have blueshifts, most galaxies show redshifts.

The shift is also referred to by the name of Doppler shift and is a phenomenon that occurs in sound waves. You will notice this while listening to the approaching and disappearing train whistle. When the train is approaching, its destination, sound waves "pile upwards" and enter your ears more frequently, as it goes away, the opposite happens. The sound waves are stretched out. Longer wavelength means lower frequency, lower pitch. For visible light, a lower frequency means a shift towards the direction of red.

The redshift value, called z is defined by the relationship:

$1 + z = n_0/n = l/l_0$

where n_0 is the frequency of light when it is produced at the point of origin (frequency that is in rest frames) and n is the observable frequency. The redshifted (or blueshifted) frequency. In the case of a redshift Z equals 0.1 that frequency is shifted 10 percent lower. The wavelength of l shifts 10% higher than the wavelength emitted(original wavelength of l0).).

Hubble Relation

Hubble could demonstrate an incredible relationship: the redshift and thus the rate of retreat from our galaxy is proportional to the distance of the galaxy being observed. If the galaxie is double as distant as it is on average, it'll have double the redshift. If it is four times as distant, your redshift is four times greater. The linear relationship between the speed of recession of galactic galaxies and distance is known as Hubble's law:

$V = H_0$

The Hubble law states that Hubble Law, H0 represents the speed that the galaxies travel at (in the units of km/sec, which is km/sec. It's also measured using the redshift) while D represents the distance (in Megaparsecs, also known as Mpc). H0 is also known by Hubble constant. Hubble constant. Megaparsec is one million parsecs, which is 3.26 millions light years. Hubble constants are kilometers per second or Mpc. The Hubble constant is not constant over time. It changes with time, however it remains constant in all directions in the present moment, as we observe space. The H0 value measured has been calculated to be 70km/sec/Mpc. If the galaxy is 10, Mpc away, then on average, it recedes in a speed of 700km/sec. If it's 100 Mpc away from Earth and on average has a speed of 7000 kilometers per second when it's 1000 miles away, it's retreating at the rate of 70,000 kilometers per second, which is 23 percent in the rate of light!

It is also the case that the parameter redshift Z is linked to the velocity in the following manner:

$V = c\, z$

Therefore, when the z value is .01 The velocity will be 3000 kilometers per second. This as well as the Hubble relation are applicable to velocities that are tiny in comparison to light speed. When the velocity is high, a more complicated formula is required. Astronomers typically refer to distances using the parameter z, which is known as the redshift.

What is the cause of this phenomena? Imagine a one-meter ruler stretched in an uniform manner to double its original size. Now imagine doing this in a single second. We put 3 ants on the ruler. Ant A at the 0.01 centimeter (cm) mark and ant B at the 20 centimeter mark, and Ant C is placed at the 50 centimeter marker (as shown in figure 2.1). Assume that the ants don't move around when we extend the ruler.

Figure 2.1 The ruler is stretched to double its length. When its length increases and all distances relative to it double in the stretching time. Since B's speed increased between marks 20 and 40 in a single second the speed of B as observed from the A side is 20cm/sec. The speed of C, as seen from A's perspective moves at 50 centimeters per second.

Imagine the ruler is set at the zero-centimeter mark (at the left side) and then it is stretched from a meter in length to two meters because of an force pulling on the right side. After just one minute of stretching how much from each other are the Ants? Ant A remains at 0cm, but Ant B is 40 cm from ant A. And Ant C is currently 100 cm from the ant A. In the perspective of ant A, ant B moved 20 centimeters in a single second, whereas Ant C moved 50 centimeters within a second. From the point of view of ant B's perspective, ant C was only 30 centimeters away prior to stretching, and 60cm away after just one second. Thus, Ant C changed 30 cm within a second from the perspective of ant B.

In summation:

B as seen from A - was 20 cmand at 40 cm 20 cm/sec

C seen from A to was at 50cm and then at 100 cm 50 cm/sec

C, as seen from B was 30 cm the distance was then 60 cm and 30 cm/sec

What can we conclude? From the perspective of each ant, other insects are moving away from the area. The amount of time they travel per unit of time is determined by the distance they started from. Their speed is based on the distances they are separated by. In reality, all separations double. Space as an entire (on the ruler) increases equally. The speed is linear to the original separation. The simple extension of a ruler shows the same pattern of behavior as Hubble observed, however this time, the Hubble stream is actually a 3D phenomenon! Replace the ants with galaxies and observe space stretching across all three directions over the span of billions of years. The explanation behind the observed redshift in galaxies, as well as that linear correlation between velocity (redshift) as well as distance that space expands equally across all directions!

Space expands uniformly across all three dimensions and has been doing so since the time that the Big Bang initiated, one can conclude through "running the film backwards" that in the past all all the energy and matter of the universe was located in the same spot. Also, it must have been extremely hot and intense because of the expansion of matter.

How long have we been seeing the expansion going for? For a reasonable approximation our universe's age can be calculated by the opposite value of the Hubble constant which is measured in units of km/s/Mpc , and this is converted to units of the inverse of time. When we calculate the Hubble constant we also measure the time to the first order by using this method.

To calculate by using the inverse of the Hubble constant is calculated as T0 = 1 / H0. In T0, the subscript "zero" indicates the present epoch. The current H0 is 70 km/s/Mpc. We can remove the kilometers and Megaparsecs that are both distance units and get H0 = 2.26 10-18 sec-1. T0 = 1. H0 is 4.42 1017 seconds which is 14 billion years. This represents the age of light travel of the Universe for light that is propagating out of at the time of the Big Bang.

It is very similar to the age that can be measured by methodologically more exact methods. Therefore, the age of the universe is approximately three times that of that of Earth as well as the Sun. Although we cannot look back towards our time in the Big Bang, we can be very close to it, through observing what's called"the Cosmic microwave background.

Cosmic Microwave Background

Although Hubble's first discovery was based upon a few galaxies, and his estimation of H0 was extremely in error (much higher than it should be, which suggests an older universe) The linear relationship between the speed of recession and distance remained verified by astronomers who took more and more measurements.

Therefore, there was ample evidence to support evidence for the Big Bang theory, but it needed further proof. A different theory, called one called the Steady State Theory, suggested that matter was continuously produced as the Universe was expanding, while the theory of the Big Bang postulated that all matter and energy was created at the time of the original event. Actually, it was the those who advocated for the Steady State theory that named the Big Bang to the expansion theory to discredit the concept!

They were, however, in the wrong place in the story. The most important confirmation of that Big Bang came in 1964 when the relic radiation leftover after that Big Bang was accidentally discovered. We are able to see this Big Bang relatively early in its evolution due to this radiation. Two of the physicists involved, Arno Penzias and Robert Wilson were working on calibration of the millimeter radio wave telescope located in New Jersey located at AT&T Bell Labs. They observed a residual 'noise signal that was identical throughout the day, and even throughout the year. This could indicate a cosmic cause that is not part of our solar system or perhaps our galaxy.

The radiation, dubbed The Cosmic Microwave Background (CMB) radiation was believed to be present in the 1940s. However, only rough estimates were available of the effective thermal temperature. Furthermore, scientists from close by Princeton University were just getting ready to begin an effort to discover this CMB when they learned of the findings that were discovered by Penzias and Wilson who were not aware of the predictions. As the discoverers, they were the first who would later be awarded the Nobel Prize for Physics.

It is a matter of discussing the temperature of radiation, since it is the so-called the black radiation of our bodies. It is the thermal radiation that is that is released by any warm body, including the Sun and the planet Jupiter as well as our bodies. The distribution of radiation based on frequency is an expression of the body's temperature in question. The hotter the body, the greater frequency characteristic in accordance with the following relation:

H Npeak = 2.82 kT

andh is Planck's constant, npeak is the frequency at which the intensity increases (peak luminosity) while it is a physical constant called Boltzmann's constant. T is the body's temperature. When the temperature increases by two and the radiation distribution is shifted to higher frequencies with the highest output at a frequency that is twice more high (wavelength shorter by 2 times). Determine the location of the peak, and you've determined the temperature in the event that it's truly blackbody radiation. Figure 2.2 illustrates the spectrum observed of the CMB and has great agreement with the black body's spectral shape.

The connection is so that for a star like that of the Sun the light is a peak in the visible wavelength, and for objects with a temperature of room, the radiation occurs in the infrared spectrum, which we can't be able to see but can sense as heat. The CMB radiation originated within the visible. It was emitted by the initial explosion of the Big Bang at a time when it was cool to about the same temperature as the Sun.

Before that at the time the flameball became more hot there was no plasma or ionized gas. It is comprised almost entirely by hydrogen and Helium, and many light particles (particles made of light) floating around. Ionized implies that all electrons were freeand not confined to orbits around their nuclei due to the fact that their temperatures and hence their energy were sufficient to allow them to remain free. Photons easily scattered off the many electrons that were free.

However, as the universe began to cool and the electrons became less active and were then attracted by the electromagnetic force from the nuclei (positive nuclei attracted positively electric electrons). This was the moment when the first atoms formed. After the electrons free of charge were attached, the gas of the fireball became translucent to the radiation. The optical radiation was no anymore scattered, and was able to travel in a fluid way throughout the universe from the time it was created.

The time this event took place at the time, the universe was approximately 400,000 years old as when compared to 14 billion years now. As the universe expands, and expanded, the CMB radiation has drastically redshifted. In the beginning, it was emitted predominantly in the optical wavelength, it now is located in the microwave part of the spectrum. its characteristic temperature of radiation is T + 2.73K. This is just a little less than 3 degrees higher than absolute zero, which is the lowest temperature that is possible. K stands for Kelvins which is the measurement employed to determine a temperature scale that starts at zero. It is Celsius (Centigrade) terminology zero degrees Celsius the freezing point of water is 273K, and temperatures in the room are around 300K. In a coincidence, the CMB's temperature CMB is nearly exactly one percent of the temperature that water is able to freeze, according to an absolute temperature scale.

The Redshift in CMB is extremely large by 1100 times! It's also proportional with the particular thermal temperature. The initialtemperature, which was not redshifted was approximately 1100 * 2.73K = 3000K. In contrast, the visible area (photosphere) is 5500K.

The CMB is the furthest back in time we could "see" that is that we can observe light. The space-time distributions of temperature and intensity of the CMB reveal it to be extremely uniform across all directions. This isotropy indicates that we are living in a universe that is homogeneous on the biggest scales. There's certainly a good amount of structure in galaxies, stars and clusters of galaxies but at the largest scales , the universe is homogeneous.

We have noticed several significant, yet tiny, changes that have led to a massive transformation in the field of cosmology in the past 10 or 20 years.

What's the matter?

The CMB radiation originates from an early stage of our universe's expansion at the time when the first atoms were formed. But what exactly were those elements? Most of the time they were helium and hydrogen that are the two heaviest elements and are by far the two elements with the highest abundance in the universe.

In the initial one millionth during the original Big Bang expansion, the universe was made up of dense and hot plasma that contained free quarks and gluons , as in addition to other particles and a lot of radiation. Matter and antimatter were plentiful[55. As the expansion continued , and it became clear that the temperatures in the universe decreased and the elementary particles that are composite known as hadrons form. There was a multitude of antihadrons and hadrons near thermal equilibrium. The majority of them were destroyed in pairs. However, when the universe was only one second old, all antihadrons were destroyed and there was an surplus of hadrons. The imbalance of matter and antimatter overcome so that only matter remained because of an excess of just one percent of 30 million, compared to antimatter. The present day antimatter is very rare in the universe.

There are two kinds of hadrons: nucleons each one composed of 3 quarks and mesons, which are composed of two quarks. Nucleons are the well-known neutrons and protons present throughout every atom. The atom that is the smallest is hydrogen, which is normal hydrogen having only one proton in its nucleus. The other nuclei of the atomic family contain at least one proton, and several neutrons along with a amount of electrons.

Within 10 seconds of the beginning at the time of the Big Bang, the Universe was dominated by powerful photons (gamma Rays). In the initial minutes, from around 3 minutes to 20 minutes old the universe remained warm enough and dense enough that nucleosynthesis took place. Nucleosynthesis is the process through the nucleons fusion together to form heavier elements like the hydrogen bomb. For instance hydrogen fusion may result in deuterium (heavy hydrogen, with two neutrons and a proton within the nucleus) as well as the second lightest element, helium (which typically has the presence of two protons as well as two neutrons inside the nucleus). The fourth and third elements beryllium and lithium were also created, but in a less significant manner.

At the conclusion of the nucleosynthesis period the universe was made by matter not non-matter comprising 75 percent hydrogen by mass, 25% helium mass. It also contained tiny quantities of heavy hydrogen lithium, and beryllium. The heavier components, such as nitrogen, carbon and oxygen, which are vital to the existence of life were formed at the time of this.

Table 2.1 The principal elements that are produced during cosmological nucleosynthesis, which was took place when the universe was only 20 mins old. Two kinds (isotopes) of hydrogen(isotopes), two kinds of Helium, and one type of lithium are created. Hydrogen contains one proton and one neutron or zero Deuterium is also referred to as heavy hydrogen. Helium contains two protons as well as at least two neutrons.

Summary

Prior to the 1980s , the general cosmology view was that the universe was expanding in a matter-dominated universe with an uncertain age ranging from 10 - 20 billion years and without certainty about the mass of the universe. The mass that was inferred from observations of stellar objects was not enough to hold the universe in a gravitational way and stop the expansion. There was evidence of gravitational effects near close to the edge of galaxy clusters as well as in clusters of galaxies and groups that additional mass was in fact present. The focus was shifting to finding additional mass that could be enough to trigger the universe to slow down its expansion in the near future. in the future.

Over the last two to three decades significant improvements in observations and theoretical advances have led to significant advancements in our understanding of cosmology. It is true that the Big Bang model has been confirmed, but it has also been modified in a variety of significant ways. These modifications are discussed in the following chapter's discussion on inflationary modifications of the Big Bang theory.

Chapter 3: Inflation

Big Bang Theory Limitations

We aren't sure exactly the exact moment when our universe began but our expanding base of knowledge on cosmology from both observational and theoretic research, has enabled us to stretch the frontier back in time to the earlier and earlier days in fact, to well within the very first fraction of one second. Also, while we aren't sure exactly how it began, we know what it looked similar to in all but the very first phases and how it has evolved from the very beginning. As we move back using what's commonly referred to as inflationary Big Bang models, we seek to determine the nature of events at very in the very beginning of time. The research aims to understand the nature of the universe as early as 10 to 37 seconds after its formation, which is under a billionth trillionth of a trillionth of an instant! The work combines high energy particle quantum gravity, physics and theories such as supersymmetry and string theory.

Although the traditional Big Bang theory as formulated in the 1920s through the 1990s has seen numerous successes, it has left open a number of unsolved issues and a few key questions. We'll review the achievements first. It is believed that the Big Bang theory is able to provide a general explanation for the development of hydrogen and Helium as being the two most plentiful elements of the universe and in the amount that is that is observed. We are aware that heavier elements like carbon oxygen, iron, and uranium are created inside the stars' interiors and we will look at this further, so the inability to produce these elements during the Big Bang is not an problem.

It is believed that the Big Bang is able to explain the expanding universe as well as the Hubble galaxy clusters that are separated from one another. If we assume that the universe was initially homogeneity, the current situation of the universe at present is a reasonable conclusion. According to the theory, the Big Bang can accommodate universes that have ages ranging from 10-20 billion years, which is generally in line with the age of the most ancient stars.

The Big Bang leaves a remnant cosmic microwave background, with a uniform temperature across the entire direction as you can see and the temperature is in line with predictions. This is the Big Bang, as it was commonly viewed in the 1990s and the 1980s believed that matter was the dominant force in the mass-energy field of the universe, as well as acceleration because of the self-gravity of the matter. Therefore, the model was simple, one of an uniform, homogeneous, and uniform entity moving apart from the moment of its creation and then slowing down in the course of its time under the influence of gravity.

The most important question to be considered was the question of whether the density average of matter was less or higher than the amount required to stop the expansion. If the density was lower then the universe could expand for ever. If it was larger the universe would collapse again - and the Hubble stream would then reverse and galaxies would begin to swarm towards each other. It was generally believed that the density varied between 10% and maybe 200% of the needed density, however there was much uncertainty.

But, to achieve an outcome where the density of today is within 10 percent of that critical amount, requires the density in earlier times to be nearly as high as the critical value with very high precision, 1/1015. What is the reason for this to be the situation?

There were other issues that came up. One of them was about how galaxies came to begin to form in the beginning. It is believed that there were certain regions with slightly greater density than others at the time they formed when the Big Bang occurred, but the smoothness of the background of cosmic microwaves suggests that the regions with a high density were different from the less dense ones by less than one part of 100,000 when the universe was only 400,000 years old. Was there enough time for these overdense regions to break free from the Hubble flow through gravitational attraction? The first galaxies appeared in the first billion years of time.

When we study the galaxies' distribution and the degree to which they are clustered or clumping, an inverse relationship, referred to by the term a "power law is seen. What is the reason for the shape of this power law determined by a spectral indice that is extremely close to 1, which corresponds the scale of invariance?

The Planck Era

The Planck scale in quantum physics refers to the point at which time and space are expected to cease to be continuous and display quantum-like characteristics. In units of time, this is around 1043 seconds, while in spatial units, about 10-33 centimeters. It is thought that the universe begun in a very compact state, without the world as we know it, and in size that corresponds roughly to Planck scale. From the very beginning, during the first Planck time of only 1043 seconds the entire universe could be contained within this tiny space of 1033 centimeters, and the Quantum description of gravity is needed to explain the state.

Because there is no confirmed quantum theory for gravity,, we aren't sure how the universe started to grow. However, physicists believe that at the Planck size, each the forces in nature were united into one force. In some way, the universe - or, in the future, the universe, expanded and then cool down, and the symmetry was broken . Gravity was separated and the remaining three forces were first. Gravitational energy is negative signs and the unifying field or force could have had positive energy. The total energy of the universe could be zero!

Four Fundamental Forces Four Fundamental Forces

Nowadays, we know we have four basic forces of nature, or the laws of Physics. They are the well-known gravitational force , the electromagnetic force, as well as the two forces of nuclear. One is called the strong force and is the force that keeps the atomic nucleus in place. The second force in the nuclear system is called the weak force and it is characterized by radioactive decay.

Force Carrier Particle

Gravity Graviton

Electromagnetic Photon

Strong Nuclear Gluons

It is weak Nuclear Weak Nuclear Z

Table 3.1 The fundamental forces and their transport particles. Gravity is described in general relativity, a nonquantum theory. There are three forces with quantum theoretical descriptions, which include electromagnetism as well as the weak and strong nuclear forces.

They manifest as fields. Each one has a specific charge particle that is part of the field. For instance it is said that the photon (light) represents the main carrier for the electromagnetic field.

Scientists are generally of the opinion that there's an additional mechanism operating called the Higgs field. This is the field (it is likely to be composed of several fields) that provides masses to particles of elementary nature. We aren't certain what the reason for why the mass of the proton and that of electrons are the way they are. In all of the Universe, proton that is at still has exactly the same weight as elsewhere in our universe. It is believed that the Higgs fields have a minimum energy that is not zero that has energy before non-Higgs particles have discovered.

While the photon, gluons along with other gluons, as well as the W and Z carriers are all observed, the graviton, nor the Higgs boson, the primary carrier for the Higgs field, has ever been observed! However , the graviton is based on solid theoretical ground, along with it is the Higgs boson (named after Peter Higgs) is strongly predicted by the Standard Model of particle physics.

It is believed that the Higgs is also the sole particle that has been predicted in the Standard Model of particle physics which hasn't been observed. (The Standard Model includes 3 of the four forces, but gravity isn't included.) It is hoped that the energies produced by the recently operating Large Hadron Collider at the CERN particle accelerator in close proximity to Geneva will be sufficient to generate Higgs bosons.

In the very beginning of the Universe, that was extremely hot and extremely dense Physics laws that we know them today did not apply. There was a single unifying field that was composed of four forces. Many scientists believe that dissociation of the first gravity force and then the subsequent strong force from the field were crucial moments in the very beginning in the evolution of our universe.

Grand Unified Theories (GUTs)

Let's now examine Grand Unified Theories (GUTs). These theories do not fulfill the claim of grand unification. However, what they are trying to do is to bring together all four forces, the electromagnetic force strong nuclear force, and the nuclear force that is weak. They don't address how gravity can be unified.

The three forces that are described as having quantum descriptions are more powerful than gravity, even at the normal energy level as well as at the energy levels of the largest particle accelerators. In higher energies it is believed that they unite to form a singular force, or field. This energy is referred to by the GUT scale, also known as the grand unification energy. It is thought to be approximately 1016 GeV = 1025 eV6. This is 1000 times less than the Planck energy of 1019 GeV equals 1028 eV.

First, we had to work on to unify the electromagnetic force and weak force. When the energy is greater than 100 and 200 GeV both forces appear to be an interaction. In recognition of their work on the unification of electrical and the weak force Abdus Salam Sheldon Glashow, and Steven Weinberg received the Nobel Prize in Physics in 1979. In the year 1977, W as well as Z bosons described by this theory were first discovered in 1983.

Vacuum Energy

The empty space is not empty. This is because of what is known as the vacuum field which is a scalar force with a value of non-zero in its smallest energy state. The quantum field theory informs us that even space itself is filled with energy and pressure. Virtual particles are able to pop out of existence and into space in the event that they draw energies from the space for a brief period of time.

Incredibly, the vacuum's pressure is negative and when placed in the equations of general relativity, it acts as an anti-gravitymechanism, which causes expansion or repulsion, rather than attraction. When the universe was first created, the universe the energy density and tension of the vacuum could be significantly higher than it is today, and the vacuum's negative pressure could have been dominant over its energy density.

Inflationary Expansion

Inflationary Cosmology was first proposed in the year 1980 from Alan Guth in order to solve some of the weaknesses and issues that were a part of the standard Big Bang model. It could be described as an extension of, and not a replacement for the Big Bang and specifically addresses the events that occurred very early in the life of the Big Bang.

Guth was originally seeking for an answer to the problem of magnetic monopoles. Magnetic monopoles are thought to be exotic particles with the magnetic charge. In the natural world, we have only observed dipolar magnets having the north pole as well as south pole. Electromagnetism's equations and quantum theory appear to allow for magnetic monopoles, if the electric charge is quantized, that is exactly what it is. If they are present magnetic monopoles are extremely massive and uncommon. The absence of any evidence of magnetic monopoles suggests their mass should be in excess of 600 GeV.

Inflation, one of the mechanisms that cause our universe's expansion to massive amounts in a short amount of period of time, can lead to magnetic monopoles that are extremely uncommon. A rapid expansion of the universe in the very early stages of the development of the universe could cause the highest homogeneity evident on the most massive scales of the distribution of galaxies as well as within the CMB. It also forces the universe towards flat topology.

Remember how the Planck time, which is the quantum granularity, or the time of the universe, is just a tiny 1043 seconds. The inflationary theory suggests that shortly after the beginning in the Universe, beginning at about 10-36 seconds it was a short period of massive inflation within the space-time fabric driven by the energy released by the field of vacuum. This could be caused by the disconnection between the force of strong and weak, and an electromagnetic force that was at the GUT energy range of 1016 GeV and could cause a massive emission of energy.

Imagine (this isn't easy to imagine!) that there exists a small space-time bubble that is merely the vacuum field that is in an energy state higher. The inflation theory is that it could be caused by a degeneration in the field of vacuum from a higher energy level to an energy state lower. When the vacuum field decreases according to the GUT theories, the strong force differs with the weak force as well as that of the electromagnetic force. The energy of the vacuum field is transformed into radiation and various matter particles. It also causes the expanding of space. Energy in general is saved, however matter is formed by the release of energy from the vacuum field when it degrades. This is in contrast to what happens during an explosion in nuclear form, where certain elements of matter are transformed to energy.

Based on the inflation theory, the universe expanded in size, starting from a size that is little more than the small Planck scale, all the way to a macro-scale. The inflation event is believed to have begun in the vicinity of 1037 or 10-36 seconds following the event, and to have concluded around the time of 1033 and 10-32 second. After this brief time the inflation process ceases and the energy released is released into energetic particles and radiation which are as the main contributor to the balance of energy. In the following years the universe is evolving much or in part according to the conventional Big Bang model, but with one significant change which is the inclusion of dark energy. It is something we'll discuss later.

When inflationary growth is occurring the space expands by doubles and then increases by a number of times maybe 100 times in the first 10 to 32 seconds of that the Universe has! It is actually hyperinflation much more than the inflation in the currency of that time in Weimar Republic in Germany. Think about the number of doubles. 2100 equals to the sum of 210 times 10. If 210 is 1024 or around 1000, the expansion factor is 1,00010 = 103 * 10 = 1030. That's 10 billion multiplied by 10 billion billion! This is an extremely rough estimate, based on the specifics of particle physics currently being worked out however, it's an enormous expansion amount. When inflation is over, the universe (what was then our visible part in the Universe) which was previously extremely small, and much smaller than a proton has increased to something about one meter in dimensions.

This kind of expansion pushes the universe to the extreme topological flatness, which is overwhelming any curvature that matter could attempt to create through gravity. This also permits an extremely homogeneous environment as the initial bubble of spacetime as well as the vacuum field would be driven towards near-uniformity during the expansion, which is how that the homogeneity in the CMB could be explained.

In addition the expansion is so large that it suggests that our universe, as we know it, is only a tiny portion of the universe made 13.8 billion years earlier. According to calculation, the area of the universe that we can see is just 10-25 to less than the universe which was created (one 10 millionth of one billionth of one billionth). It's like if our part of the universe were one molecule of water in a liter of water!

From now on when we talk about the universe, we'll generally refer to our small space in the universe that we can observe. The rest lies beyond our reach. Because it is moving away from us at a faster rate than light it is impossible to be able to see it.

What is the possibility that other regions of the universe could escape us faster that light speeds? Relativity is a limitation on particles and matter, including photons to move the speed of light. Space itself can stretch and actually does stretch with much more speed that light speeds.

The theory of inflation was created in the late 1980s and early 1990s through research by Guth and others, it swiftly gained popularity with scientists studying cosmology. There were several technical issues to be resolved and also the question of additional observational evidence for inflation.

Dark Energy, Dark Matter

At the end in the second half of 20th century, there was a significant shift in cosmology observation. We have evidence from both cosmic microwave background (CMB) observations (that is that, the observed very small spatial inhomogeneities that occur in the CMB) as well as from further distances Type Ia supernovae[7] (measuring absolute brightness as compared to. redshift) which indicate it is flat topologically as it continues to grow for a long time. The most interesting result is that the density of mass energy in the universe is controlled by first and foremost what called dark energy (which makes up approximately 73 percent in the entire) and, secondarily by darkness (about 23 percent). Light-colored matter, also known as normal baryonic matter, makes up less than 4 percent of the total mass-energy density. This means that there's nearly six times the amount of dark matter as normal matter, and more than three times the amount of dark-energy as that of the dark matter (in equivalent units when the formula $E = mc2$).

The inflationary models say that the total of these three components will be able to be equal to the so-called critical density. That's exactly what's been observed to quite high precision, in the context of the statistical uncertainty of the tests. The critical density is based solely of the Hubble constant as well as the the constant of gravitational force. If the balance between mass and energy was totally dominated by matter having a density higher than this threshold would mean that the universe will eventually collapse because of its own gravity. Because the dark energy dominates the mass-energy balance, it cannot be able to recollapse, and it will continue to expand.

The model that is most consistent in explaining CMB observations and supernovae Type Ia observations and observations of how galaxies are clustered as well as the abundance of deuterium and helium in the universe, as well as other cosmic tests is known as the LCDM Cosmological Model orLambda which is the Cold Dark Matter model.L(Lambda) is also known as the cosmic constant that Einstein predicted and later was later rejected. Einstein added it to the static model of cosmology that was constructed before the time it was discovered that the universe was expanding. When it was discovered that the Hubble expansion was discovered, Einstein dropped the word, calling it "his biggest mistake". Actually, it appears the Hubble expansion was correct in incorporating the cosmological constant, however because of the wrong reasons.

The cosmological constant is understood as theorrent amount of energy density in the vacuum. A quantum mechanical sea made up of particles within the vacuum creates an "negative pressure" that is precisely what is required in order to understand the mysterious dark energy concept which is the cause of an expansion. The current expansion isn't quite as big as the one during the inflationary era but it is significant enough to control the dynamics of our time and in the near future.

The period of inflation in the early stages of the universe's development can be understood as a time in which the vacuum energy was enormous. When inflation was over, it would have slowed down to a smaller amount which is in line with the current state of affairs known as the dark energy or cosmic constant. Particle physics doesn't yet provide a convincing explanation of what the L value is, whether during the period of inflation or at present. The current value is lower than what could be predicted by general argument.

The Summary of History of the Universe

After the period of inflation, the expansion continued but at a much slower pace. In the first second, the temperature had dropped enough that neutrons, protons and electrons were the primary components of matter, and the era of matter-antimatter annihilation was over. But, the mass-energy was mostly photons until about 70 000 years following the Big Bang. Then, dark matter is dominant[8] and regions that are a little over-dense could begin to develop into larger structures with the help of gravity.

The background of cosmic microwaves dates to 380,000 years following the start of the universe. The first galaxies and stars emerge in those first few billion years when dense regions collapsing gravitationally. In the initial 9 billion years or more dark matter dominates the balance of mass and energy in the universe. This is the time of galaxies that have redshifts that are approximately 1/2 . However, as the space volume increases as a result of the Hubble expansion, the dark matter density decreases and the dark energy is the main component that is responsible for around 3/4 of the mass-energy in the universe. The expansion was slowing down due to gravity of matter's self-gravity in the beginning of the universe's existence. When dark energy was dominant it began to accelerateand not slow down. The current condition of the universe has galaxies growing away from one another at an greater and greater speed.

We will look at the dark energies and dark matter in more detail in the subsequent two chapters, starting in dark matter.

Chapter 4: Dark Matter

What is it that makes it Dark Matter?

The idea that there was a "missing mass" or "dark matter" was among the first ideas scientists and cosmologists to grapple with while they were constructing the Big Bang model was being established. The concept of dark matter was thrown a topic as a central aspect in cosmology and astrophysics since the 1930s. Fritz Zwicky at the California Institute of Technology was studying the internal velocities of galaxy clusters. The clusters of galaxies are gravitationally linked thing that contains hundreds, or even thousands of galaxies. Gravity mass for the galaxy can be identified by studying the velocity of various galaxies that are part of the cluster. He noted there was a dispersion in velocity (variations) within dense clusters of galaxies revealed that the masses of the cluster were 10 times more than the expected light contributions of these galaxies as well as the mass-to light ratios were popular in the time.

This method shows how much gravitationally bound mass in the cluster than could be determined by the visible light of the galaxies by themselves. This conclusion is further supported by the most recent measurements on the temperatures of gas within the clusters, which is so hot that it produces radiation called X-rays. The greater the mass of clusters, the more the heat generated by the gas by frictional processes when it is placed within the gravitational and wells of the cluster.

So this is how the "missing mass" problem was created and more mass was deduced by gravitational forces than could be directly observed in other wavelengths, or in visible light. What we call today dark matter was being observed by gravitational force, but was originally referred to in the past as "missing mass" due to the fact that it was not emitting light.

In addition, evidence developed over the course of time of the existence of flat "rotation curvatures" (which are used to measure the speed of rotation in relation to. distance from the center) in galaxies. This suggests a that the matter distribution is much more extensive that the distribution of light. The process of creating a curve for rotation is through measuring orbital speed of gas and stars in various locations away from the central point of the galaxy. For a galaxy with a disk you would have to extend outwards from the center of the disk on the surface that runs through the disc. The observed speed of rotation is dependent on the square root of the mass's inside to the point of interest.

At first, one might have thought that once the light faded away (no additional visible matter that is significant beyond a certain date) that the speed measured will then reduce. In reality, it was discovered that spiral galaxies generally have flat-rotation curves, as illustrated in Figure 4.1 and their mass-to-light ratios increase as you move farther away from its center. The velocity within the regions that are in outer space is approximately similar to that of areas in the middle. Halos of the galaxy, also known as the outer regions, contain material that is difficult to be detected.

Figure 4.1 The typical galaxy curvatures of rotation look like "B" (Credit: Phil Hibbs)

Both closed and open Universes

The universe is expanding ever since beginning of the universe, which was around 14 billion years earlier. We also recognize that gravity acts as an appealing force that attracts all particles with weight (or the energy). This means that all matter has been pulling on of the other particles throughout history of the universe and causing a slowdown in the expansion. So the question is, does gravity slow down the expansion of the universe enough so to allow it to be able to reverse course and return towards its original form?

It's conceptually like firing an rocket. Does it have enough energy to reach the velocity of escape (determined according to the power of earth's gravity) and crash into Earth?

For a homogeneous matter-dominated universe, it is only dependent on the degree to which the average density of the matter called r (Greek word"rho) has greater or less than the critical value. The value is based upon G which is the constant of gravitation, and on the area of the Hubble constant H0 as per the following relationship.

8p Grcrit = 3 H02

In other words, for a density that is lower than that of the crucial value the universe continues to expand for ever and for r>r crit the universe it collapses at some time in the future dependent on the precise value ofr. We refer to the universe that is constantly expanding as open, while the one that collapses to itself closed. This is a reference to curvature of space-time too, taking into account the universe in its entirety in both scenarios. We refer to these as Friedmann models of universes that have curvature that is either positive (closed) and positive (open) curvature. The situation where r = rcrit is referred to as the flat universe. It symbolizes the boundary between two types of closed and open.

Omega

To make it easier, cosmologists have defined an attribute that relates between the density of the median and the critical (using omega as the Greek word omega).

O = R/rcrit

In this sense, when O 1.01 the universe is wide open and curving in a concave way and continues to expand forever. When O > 1, it is considered to be closed, and is curving in a convex manner similar to the way that the Earth's spherical surface is curled and will eventually be able to collapse back into itself in the future. If O > 1, it is theoretically flat and eventually it stops expanding however it does not completely fall back to itself.

The invisible matter is known as "missing" due to the fact that we are unable to observe enough normal (luminous) material to completely close the universe. Cosmologists believed that a universe with O > 1 or maybe higher was visually preferredsince it would cause a collapse and a possibly a Big Bounce or Rebirth to the world. However, there is no scientific reason to believe that it was necessary to make this the scenario. The most important thing isn't the philosophical value of theories, which are subjective, but rather whether it is able to stand up to the facts.

In actuality the universe that is dominated by matter and O > 1 has a lower age and is less than 2/3 in the Hubble duration (1/H0) that is 14 billion years. Moreover, it is hard to compare this age with the age of the oldest stars. As the observations regarding Hubble constants have been improved, and the measurement of Hubble constant have enhanced, the measured value has increased from 50 to 70 km/sec/Megaparsec. With the Hubble constant of 70 as it is now the cosmology that is dominated by matter with O = 1 is a time-scale of just nine billion years. Some of the most ancient stars have been to be based on the simulation of their evolution to be more than 10 billion years old There were signs that a model that is dominated by matter and has O > 1 is not ideal, and therefore evidence suggested that the universe is open.

The Revolution that started in 1998

There was growing evidence during the second half into the second half of the century of an important contribution by "missing mass" also known as dark matter principally through the analysis of gravitational effects on the mass. It was believed to contribute significantly greater to average density, than normal luminescent (baryonic) matter.

In the mid-1990s, it was believed that dark matter is the main component of the universe, it didn't seem to be sufficient amounts of dark matter for it to "close" our universe or render the universe topologically level. It appeared that there was just a quarter or a third of the density needed for closure, based upon these two formulas as well as the value that was measured of the Hubble constant. This would have allowed for an age of the universe that was in line with the maximum stellar ages due to the wide range of uncertainties at the time of the value measured of the Hubble constant. We know the dark matter to be crucial and has a greater influence over ordinary matter. However, it appears that in the balance between mass and energy in the Universe, something known as dark energy is more crucial.

In the last ten years, it has witnessed a significant shift in the field of cosmology due to amazing observations made in the optical and microwave areas of electromagnetic spectrum. There is now evidence of the cosmic microwave background (CMB) observations (that is, the spatial distribution of inhomogeneities) and also from far-off Type Ia supernovae (luminosity vs. redshift) which suggest that the density of mass energy in the universe is controlled in the first place by the dark energies (about 73 percent) and then, in turn through darkness (about 23 approximately 23 percent). The luminous matter, which is common baryonic matter that comprises stars and gas in galaxies, accounts for only 4 percent of the overall mass-energy density.

The percentages are normalized according to the current critical density 1029 grams per cubic centimeter, in mass units, or 10-8 erg/cc in energy units with the formula $E = mc^2$. The universe is extremely empty. If it was made only of normal matter, it could be as large as one hydrogen atom per 200 Liters of space! The three components are illustrated in the figure below, which is a pyramid-like shape.

the masses-energy equivalents that are observed from ordinary matter as well as dark matter, and dark energy. Ordinary matter constitutes a small portion of the total mass-energy within the universe.

The high-resolution measurements of CMB from COBE Boomerang, and in particular taken from the WMAP satellite, permit us to analyze how much power is absorbed by temperature variation at the angular scale.

The CMB anisotropy spectrum shows several peaks in angular scales from 1deg to a few millimeters. Three distinct peaks are evident as well as the magnitude and location that these "acoustical" harmonics of the reflection radiation (due to the fluctuations in the plasma during the moment of CMB dissociation) this allows us to very precisely determine the cosmic parameters. Combining the CMB observations with data that comes from Type Ia supernovae luminosity vs. redshift as well as from observed spatial correlations between galaxies (degree of clustering into groups or clusters) Cosmologists seem to have verified the fundamental specifics in the inflationary Big Bang model.

The chapter will focus on dark energy. we'll not go into further detail about the dark energy aspect and will leave that topic to chapter 5. This chapter will only highlight the necessity for dark energy could be the reason Einstein included the Cosmological Constant in the equations of general relativity. This article focuses specifically on the dark part of matter that comprises approximately one-quarter of mass-energy.

What are the possibilities? Types of Dark Matter?

Dark matter can be barren (ordinary matter) or non-baryonic. Non-baryonic matter can be "hot" that is, it's made up of relativistic, light particles as well as "cold" and consists of massive non-relativistic, non-relativistic particle (axions might be an exception, as light particles that are not relativistic). The non-baryonic classes are also known by the names "hot dark matter" or "cold dark matter".

Baryonic

Dark matter from baryonic sources could make up 10% to 15% of overall dark matter (i.e. approximately 4% of total mass-energy) however, the evidence suggests it is not able to explain all the darkness. For instance the hot gas that is present that is found in clusters of galaxies releases radiation and can account for at least 20 percent of the mass binding regardless of whether it is the result of the fall of primordial material or is ejected from galaxy's constituent galaxies. The X-ray temperature of a cluster is a reflection of the total dark and visible binding mass of galaxy clusters. The temperature values seen in high redshift samples match the cosmological parameters previously mentioned that, which include the energy and mass contribution shown in figure 4.2.

Neutral hydrogen is detectable in the radio spectrum at 21 cm, however it isn't present in sufficient amounts to account for just 20 percent of the mass derived through galactic orbit curves.

In the intergalactic space, the primordial gas would be dominated by hydrogen and could appear by absorption lines cold, or with the X-ray or ultraviolet spectrum when it is hot. The Gunn-Peterson phenomenon would be the recognition of an absorption line in that blue color of the Lyman the alpha line within ultraviolet. This trough cannot be seen in quasars with high redshifts and suggests that the intergalactic medium has been being reionized in the early stages (by redshifts $z > 20$) due to ultraviolet radiation from the galaxies that were first formed and the stars. Neutral hydrogen residual is seen in the absorption narrow Lyman Alpha "forest clouds" observed against quasars however, these clouds may amount to less than 10% of the total baryonic matter.

The most sought-after barsyonic possibilities are known as MACHOs. They are massive compact halo objects which may be faint white dwarf stars, brown dwarf stars, or even primordial black holes created in the Big Bang. The faint red stars can be detected through stacking a variety of galaxy images using broadband filters, but they are not considered to be the primary cause of these observations. Primordial black holes (PBHs) are believed to have formed prior to in the Big Bang nucleosynthesis era, and they would have to be greater than 10^{12} kilograms to be unable to radiate away through Hawking radiation in the time of the present universe. The most reliable cosmological models seem to eliminate a sufficient number in black holes. On the other hand PBHs could serve as seeds for central black hole formation in the galaxy by accreting dark matter within the range between z 1100 and the z-value of 30.

Brown dwarfs, weighing about 0.1 solar masses, still remain an attractive chance to determine the predominant baryonic component. The baryonic component is not able to explain the majority of dark matter, but. There are some limits to this component due to deuterium's abundance. The majority of the deuterium that is observed throughout the Universe was produced in Big Bang nucleosynthesis in the initial few minutes.

In stellar nucleosynthesis deuterium is used to produce helium however, during Big Bang nucleosynthesis the universe expands and cools prior to the conversion between deuterium and helium is be complete. The amount of deuterium from the beginning limits the volume of baryonic material to approximately 4 percent of the mass-energy observed in the universe. This is based upon the Hubble constant that is believed to be at or near 70 km/s/Mpc. This means that baryonic matter only comprises approximately 1/6 of the mass observed.

Then we are left with the fact that 5/6 of the material in our universe are not baryonic.

Hot dark matter

One of the main candidates for dark matter that is hot has been the neutrino that is actually a family of non-charged, low mass particles. While neutrinos are believed to be plentiful in the beginning of the universe, relativistic neutrinos that formed during the Big Bang cannot be the most important Dark

Matter source. If they were to dominate the matter part of the universe, they would be dispersed to permit clustering in a sufficient amount to recreate the observed populations of galaxies as well as the observed filamentary and clustering structure.

The phenomenon of thermal damping is seen when the hotness of dark matter (e.g. neutrinos) on an extent of 4x 1015 solar masses. It is bigger (about 10,000x) than the average galaxy mass, meaning that no galaxy seeds are able to develop. Its damping spectrum spatially extends up to approximately 25 Megaparsecs. At this magnitude, gravitational collapse will only be observed in the present time or even later, but we actually see galaxies as well as quasars that date back to an age that was young, z 7.

Cold dark matter

Axions could be a dark matter that is cold (CDM) potential candidate. Although their masses are thought to be tiny and non-thermally produced, they have been observed within the first universe therefore, they are not relative. The experiments at Lawrence Livermore using low-noise electronic amplifiers as well as the Kyoto Observatory Kyoto using high excitation Atomic state (Rydberg Atoms) have eliminated axions from particular mass levels (of only a few microeV) as the primary components of a galactic dark matter-dominated halo. However, axions that have different masses are still an possibility.

WIMPs which are massively weakly interfering particles, are the most likely candidates for dark matter that is cold. Because their masses are extremely high, the particles are extremely slow-moving and non-relativistic. They are, therefore cold. WIMPs are only interacting through gravitational force and weak nuclear force, and therefore are difficult to spot. WIMP masses are believed to be at least 10 GeV.

Based on CDM as the primary matter component, the densities fluctuations throughout the Universe's beginning may expand by taking an "bottom up" method. This permits the formation of large-scale clusters (masses of 105-106 solar masses) followed by galaxies, followed by clusters of clusters and galaxies, and even superclusters, too. The 25 Megaparsec scale is corresponding to an upper limit for the clumping process and overdensity (gravitational divergence from normal huge-scale Hubble flow) and not the lower limit.

One of the most probable candidates to be used for WIMPs include supersymmetric particle. Supersymmetry (SUSY) is implied by the unification of weak and strong electromagnetic forces and electromagnetisms at high energies. It is also suggested by the string theory. In SUSY it is believed that the Standard Model particles have partners with spins that differ by half (thus the bosons are the boson partners of fermions and the boson partners are bosons is fermions). Potential heavy particle (WIMPs) comprise the phototino (photon partner) as well as the zino (Z boson's boson's companion) as well as the higgsino (partner of Higgs boson, that's not yet known! However, it is it is widely thought to be present). It is important to remember that SUSY was created naturally from physical considerations for particle physics, not out of a desire for explanation of dark matter but the stable SUSY particles could provide an explanation using the appropriate quantity of material.

The heaviest SUSY particle, whatever you choose to call it, can be believed to be stable and because the particle is non-neutral and interacts in a weak way, it's an excellent candidate for the main element of CDM. In certain models, the neutralino is a superposition that combines the photino, and higgsino states, is considered to be the most light supersymmetric particle which is why it would be the most straightforward to identify.

Research and Development Detection and Limits

This suggests the not-baryonic CDM will be the major source of mass within the Universe. The mass of the particle for CDM is significantly higher than 1 GeV which is around 100 GeV. There is a possibility of detecting dark matter, however these are only preliminary, and dark matter is observed in indirect ways from observed impacts of gravitation (which are galactic rotating curves, velocity variations in galaxies that are clustered as well as gravity lensing). The brand-new Large Hadron Collider at CERN is now operating with colliding energies of up to fourteen TeV (about one-quarter of the mass of the rest proton) and could offer direct detection of supersymmetrically symmetric particle partners.

Cosmic WIMPs can be identified either through indirect or direct detection. Direct detection can result through recoil caused by elastic scattering. Direct detection is possible after the destruction of WIMPs in the galactic halo or the Earth or Sun and the subsequent discovery of the decay product which could include neutrinos muons, gamma rays, or antimatter. Theoretical arguments and some observations suggest masses that range from 10 to possibly 500 GeV.

There could be a signal from the DAMA/NaI collaborative (based on Gran Sasso National Laboratory in Italy) of nuclear recoil that was measured using 100kg of NaI(Tl) (thallium-doped sodium Iodide) scintillators over a seven year time frame. These tests depend on an annual recurrence caused by the Earth's orbit of a WIMP wind that is believed to be. The signal is found to be the highest during June, when Earth's orbital velocity in relation to the Sun is the closest alignment with the Sun's orbital path in the Milky Way. If the signal is indeed it is due to the scattering of dark matter the observed cross section corresponds to theories of neutrino-dark matter within "low-energy" SUSY models. A new 250 kg DAMA/LIBRA detector that is now available has increased the detection level to a high 9-standard deviation level of confidence.

A large amount of diffuse gamma-rays with a wavelength of more than one GeV of the EGRET satellite could indicate an indirect sign of neutralino destruction in the halo of our galaxy. The reason for this may be due to normal cosmic sources of gamma-rays however (e.g. gamma ray bursters). Recently, Hooper and Goodenough have reported a discovery of dark matter annihilation and transformation into tau leptons, which is approximately 1o away from the galactic center with the Fermi Gamma-ray Space Telescope. This is the region that lies within 600-light years in the galactic center, just a few degrees away from the massive black hole in the middle of the Milky Way galaxy, and therefore a region with a significantly increased density of matter. The researchers observe an increase in the signal in gamma rays with energy between 2 and 4 GeV and suggest that there is a value of 8. GeV for the mass of dark matter particles more on the lower end of the spectrum.

A novel method to determine the distribution and size of the presence of dark matter via measurements of gravitational lensing in a statistical manner. In the galaxy surveys conducted by an Hubble telescope that is part of the COSMOS project, the shapes of half million galaxies were combined and averaged, with their respective distortions were used to estimate the distribution of dark matter in the space between. What was mapped by the statistical tours de force, which included XMM-Newton Xray satellite observations of galaxy clusters and optical redshifts from the ground of the survey field it is a filamentary system which is consistent with both optical galaxy surveys as well as galaxy clustering and grouping simulations.

Another project that is designed to detect darkness directly, is called the Alpha Magnetic Spectrometer, to be lifted up to space by the International Space Station in early 2011.

Summary

Dark matter is similar to the concealed part of an iceberg that lies beneath the waterline.

The part that is hidden is the main part of the mass that is the one that supports the structure visible from the surface. Dark matter is linked to normal matter by the force of gravitation. The visible matter that we are familiar with that we observe by the light of galaxies and stars, is similar to the area of an iceberg over the waterline.

Why is dark matter important? It is the dominant factor in the mass-energy density of the universe in the beginning of its life. After the end that is known as the background of cosmic microwaves,, the universe is dominated by darkness (dark energy is dominant significantly later). However, there is also ordinary matter, which at the moment is a very homogeneous gas made up consisting of helium and hydrogen Atoms, however it contains slight underdense and overdense regions.

The presence of a significant amount of dark matter causes the more efficient collapse of gravitational gravity of overdense regions. This is a self-gravitational mechanism where regions that are slightly more dense that the crucial mass density (which corresponds to the standard masses of our universe) at any point will begin to collapse from the general expansion that continues to surround them. The dark and normal matter within a particular region collide, but it is the normal matter that creates the first stars as it is interacting with various physical processes (think radiation, friction and so on.) in a greater extent. The dark matter, interfering solely through gravity, is spread out and more diffuse. Dark matter aids in the process of collapse by expanding the self-gravity of the area, resulting in the formation of galaxies and stars. There are galaxies and stars that have created in the early stages than could be expected in the absence of significant dark matter.

From cosmological observations , such as the CMB and high-z supernovae we have discovered that dark matter accounts for around 1/4 of the mass-energy amount of our universe. Dark matter is made up of either ordinary matter that is faint or most likely exotic matter that is interacting with gravity and weak forces alone. Dark matter is easily detected through its gravitational effects on the galaxy's rotation curves and can be inferred from thermal kinematics as well as the X-ray temperatures of galaxies clusters.

Dark matter is also observed through gravitational lensing processes that occur in the galactic halo as well as in cosmological structures that are very large-scale.

The high concentration of deuterium derived from Big Bang nucleosynthesis severely constrains the baryon count (amount of normal nucleonic matter) in the universe. It also can be concluded that 90% of the matter is not baryonic. This is because the MACHO contribution is small. However, the WIMP contribution is predominant. The hot WIMPs (e.g. neutrinos) are excluded due to their inability to cause clumpiness and creation of galaxies in the beginning of the universe. Dark matter that is cold and non-baryonic is the best candidate with the most popular candidate being the lighter of the SUSY particles (LSP) because it is stable. Neutralinos are thought to be the ideal choice for LSP with masses of around 50-250 GeV.

It is possible to detect direct evidence of an annual variation in the WIMP wind in a massive scintillation array. There is also the possibility of indirect detection in the form of an excess of 1 GeV the gamma radiations that are found in the galactic Halo. More sensitive detectors are required to detect these mysterious particles, and also to create an even stronger foundation for supersymmetric physical physics. The next decade will let us shed new illumination about dark matter.

Chapter 5: Dark Energy

Why does it happen? Dark Energy?

Dark energy or dark matter, at a minimum through the use of the constant cosmological as described in Einstein's equations of field, has been a possible theoretical concept from the beginning of the general relativity formulation. It is still mysterious in many ways. We do not know what it's made of or the reason for the importance it gives it. But, dark energy dominates the universe, accounting for 70% of the mass energy balance in the universe, according to the results of both cosmic microwave background (CMB) and research on the distant Type Ia supernovae. It is more powerful than ordinary matter however, it also overwhelms the dark matter discussed in the preceding chapter. That has been the situation over the past billion years.

Temperature fluctuations less than 1/10 of a degree are observed in the background of cosmic microwaves that is measured by the Wilkinson microwave Anisotropy Probe satellite (Credit NASA/WMAP Science Team)

The parameters that are influenced by CMB outcomes include, among them the time span of universe (13.8 billion years) as well as the topology (flat) and the mass-energy density in total (about 10 to 29 grams for each cubic centimeter, or 10-8 erg/cc) and the spectral index of initial fluctuations in density (n = .97 which is near to one). The mass-energy density overall appears almost similar to that which is the critical, and this results in flat topology, as suggested in inflation theory. Big Bang model. Surprisingly out of these CMB observations, we discover that dark energy has been as it has been many billions of years, the most dominant component of the universe. It is being followed by dark matter and then with ordinary matter. Because dark energy functions as an absorbing pressure-like force it can be responsible for the growth in the Universe.

In chapter 4 , we have discussed the dark matter element, which, just like dark energy, isn't completely understood. We will focus here on the most apparent component of the universe's mass-energy that is dark energy. We look at dark matter within the context of our discussion on dark energy. The theories of particle physics don't provide an adequate explanation of what the significance of dark energy could be. It has been observed to be much less than what would be expected from the general quantum physics argument.

Because the nature of the Universe has been seen as flat which means that the mass-energy content is exactly equal to its critical density. We could write an equation that normalizes mass-energy content in relation to the critical density.

1 =Ob +Odm+OL

Each equation is the proportional part of the critical density that corresponds to approximately 10-29 gm/cc matter or equivalent energy. Cosmologists consider what proportion of the mass-energy of the universe exists in normal (baryonic) matterOb. They also consider which proportion is found in dark matterOdm and what percentage is of dark energyOL. Dark energy can be represented using the Greek letter L (capital lambda). This is the term used to refer to the Cosmological constant. Because there isn't any other major contributor to mass-energy, and in the flat topology observed, the three terms need to sum to 1 with units representing what is known as the critical density. Critical density is proportional the size of the Hubble constant which is an indicator of the speed that the universe expands. Universe's age is equal to the opposite that of Hubble constant. Hubble constant.

How do we define Dark Energy?

In the 1980s, and even into the 1990s, the dominant belief among cosmologists was that the universe was heavily dominated by matter. However, at the same time evidence was growing in the sense that normal material and dark matter is not as high as the critical density which is insufficient to slow down from the Hubble expansion. The inflationary model was becoming more convincing to cosmologists since it resolved critical problems that were not solved by the inflationary Big Bang model. It predicted flat topology, with the mass-energy density being exactly equal to the critical density. This is what's been observed. Remember that $E = mc^2$ and it's the combination of energy density and mass. Additionally, in accordance with general relativity it's not just the energy and mass which matter as well as the stress and pressure strain (viscosity) contribution.

Therefore, one would require more mass-energy in order to exist in any form within the universe. Evidence for flat topology (with density that was equal to critical value) was growing through observations of the galaxies' spatial distribution (clustering behavior) as well as from the CMB. At the end of the millennium better results for distant supernovae luminosity as well as the details of the tiny spatial changes observed by the CMB temperature showed in an amazing way how the expanding universe is growing faster and faster! These findings established the obvious necessity for a significant dark energy component within the scenario in the inflationary Big Bang models.

Dark energy is compatible with general relativity and has the peculiar characteristic that it is a negative force. It behaves in a different way to ordinary material or the dark realm that are self-attracting by gravity. Dark energy is a repulsive force and is and is "anti-gravity" kind of thing. Gravity produced by matter, be it ordinary or dark matter acts as a stopper upon the growth of space. Dark energy functions in the opposite way as an accelerator, expanding the space-time fabric.

It is evident that the dark and the dark matter balance has changed over time. The development of structure in the universe , such as galaxies, stars and planets, as well as living organisms, require gravity-driven clumping of matter on various scales. This occurs in the matter-dominated phase. If the universe were controlled by dark energy for the duration of its lifespan it would be impossible to create the kind of structures we see today would not have formed. In the initial nine billion or more years, matter was dominant, but the expansion was slowed considerably and structures developed at various dimensions.

However, as matter has become less dense because of the expanding universe, dark energy has emerged as the main aspect. Over the last five billion or more years, the dark energy has been dominant over dark matter, and the accelerator is now more powerful that the brake. We can measure the contribution of dark energy but we do not know what dark energy actually is. It could be due to the energy from the vacuum state or due to the creation of a new scalar field.

Pure vacuums do not possess zero energy. there is an irreducible amount of energy because of an "sea" composed of virtual particles that pop out and back at very small intervals. Virtual particles aren't mathematical constructs that are not real and have real-world effects. For instance, Stephen Hawking showed that black holes emit light when the virtual particle pair manifests on the black hole's boundary. One of the members of a newly-created virtual particle pair could be able to escape the boundary while the other is reabsorbed into the dark hole. Another proof of the existence of virtual particles is shown by the fact that two metallic plates uncharged are placed close to each other in an atmosphere. They are attracted in part to the quantum character of virtual particles. This is observed and is called Casimir effect. Casimir effect.

Two different theories of dark energy include the cosmological constant as well as the quintessence. The cosmological constant is constant across time and space, whereas the quintessence model would change in time.

There are various forms of the equation of state which reveals the relationship between density and pressure for dark energy.

The nature of dark energy is identified using the equation of state that connects the density to the pressure in accordance with the equation (pressure is equal to density times squares of the light speed)

P = Wr C2

Normal matter is composed of positive numbers for w. In order to create an accelerating universe, the general relativity demands the equation of dark energy state to have a negative coefficient that is satisfying

We w 1/3

For a cosmological constant that is uniform and inert with time, it is w = 1. For the quintessence type of dark energy, the formula is -1 the value of w is 0. The greater the negative value of w is, the more repellent darkness is. Quintessence changes over time and usually varies within space. There have been theories suggested that the quintessence value is extremely high in the initial period of inflation and then falls to a low level to explain the current power of dark energy. This attempts to understand both the initial inflationary phase as well as the more prolonged, steady and recent acceleration phase result of the same field shifting between different phases.

Experimental Detection of Limits and Limitations on W

The most recent measurements from WMAP as well as clustering observations taken from the Sloan Digital Sky Survey of galaxies reveal that the state equation parameter is -1.1 with a normal deviation of 0.14 (68 percent certainty to be within between -0.96 > W > -1.24) which is in line with a model of cosmological constants of w = 1.

The most recent Type Ia supernovae data combined with the data above limit w to -0.98 with a single normal deviation of 0.05 and 68% confidence of being within that range -0.93 + w = -1.03.

Other methods for finding or limiting the equation of state are gravitational lensing research and the measurement of the number of galaxies in clusters in relation to the redshift distance. Gravitational lensing happens when a primary galaxy or cluster bends light coming from the background galaxy because of gravity which is the general relativity model predicts. The amount of clusters that are rich of galaxies in relation to of the redshift (distance) is contingent on the specifics that are part of the cosmic model, including the w.

Although these tests aren't a way to exclude models based on quantum mechanics however, they significantly limit the variation of dark energy as time passes and suggest that the more straightforward explanation of a cosmological constant for an unchanging dark energy remains the most straightforward explanation of the different scientific tests on cosmology.

Runaway Universe

Because the dark energy is the dominant force over all matter, the universe has entered an exponential acceleration. It is now getting closer to what is known as"the de Sitter solution to General Relativity equations. As the universe is closer to the de Sitter condition , then scaling factors will increase exponentially over time.

A = A0 exp(Ht)

And a0 represents the scale, or relative size of the universe of today. the time t while H represents an approximate Hubble fixed value in a specific date. This is an exponential expression of the product Ht, and the scale factor increases rapidly and without limits. In the universe of de Sitter H remains constant in both space and time. (In the case of the more general it is the case that it is the case that Hubble constant is only constant within space.) This Hubble constant will diminish in the near future, and be smaller than what it is now. After another 30 to 40 billion years of time have gone by, it will have frozen and remain at a constant value of 60 km/s/Mpc or approximately 14% less than the actual value which stands at 70.

The universe will then expand by doubling in size every 11.3 billion years (9)! This means that after an additional 113 billion years when it's at 127 billion years old the universe will have seen almost 10 times the size of doubling as well as the scale of every dimension of space will be around 1000 times bigger than the present! The galaxies are expected to be about 1000 times further from each other than they are currently. The magnitude will be between 10003 and 1 billion times greater than currently.

After just one trillion years the number of doublings will reach 90 doubles, and the size of each dimension will exceed 1027, or one billion billion times bigger. The size will be 1081 times greater. This assumes an cosmological constant that maintains the value that is currently observed over the 1 trillion year time frame and is speculation from our perspective.

Our horizon, or what part of the universe can be seen by telescopes, will continue to shrink in relation to. A galaxy with percent or less of its speed recession that is visible to our current location, approximately 3000 km/sec, is at a distance of 43 Mpc and H0 of 70. The galaxy will disappear out of our view once the universe has expanded seven times its size approximately 79 billion years in the future. The distance it travels will be 128 times larger and its distance to us is more than the speed of light. There will be any method to discover or connect with the distant galaxy. Keep in mind that space itself could grow faster than light.

As time goes by, less and fewer galaxies present to us today are able to be detected. They will no longer form an integral part within "our" world. As they fade from view , their luminosity will be heavily redshifted into the infrared, and later microwave, and finally radio parts of the spectrum, exactly like the CMB is redshifted heavily. They will then become inaccessible. The only galaxies that could be seen from our position are those to whom our gravitational bonds are within the Local Group. Local Group.

Summary

Why is dark energy so important? Five billion years ago and to the future indefinitely it is the dominant energy content of the universe. It triggers a reacceleration the universe, and consequently blocks the further clustering of galaxies on massive dimensions. It also hinders the collapse ("Big Crack") of the entire universe, and even significant parts that are part of it. As long as dark energy holds its positive impact into the future and extends the lifespan that the universe has to billions of years, or more - much more than what could happen should the universe be controlled by matter alone and having a density of the critical value or higher. Dark energy is thus able to increase the time available for life to develop and develop on the planets that are found throughout the universe.

From observations of cosmology, including those of the CMB Supernovae with high-z spectra we can see that dark energy is the main component with around 3/4 of the mass energy amount of matter in the universe. Dark energy is characterized by as a negative pressure and it is responsible for an acceleration at the rate of expansion. This is the case regardless of whether it's because of an cosmic constant (vacuum energy) or quintessence. Many observations are beginning to define the equation of the state parameter w, which will allow us to decide if the model of cosmological constants or the quintessence model is more appropriate to the.

A new initiative being conducted to better examine the importance to the equation that determines the state parameter w as well as how it changes over duration is called the

Dark Energy Survey. The DES will begin operation throughout the southern hemisphere at the end of 2012 and employ four different methods to further limit the parameter w. These include the measurement of supernovae, galaxy clusters as well as gravitational lens measurements, and measurements of galaxy distribution on very massive scales. The most massive acoustic changes throughout the universe were observed at the time that neutral hydrogen first created, and at that time CMB was released. The fluctuations created an impression on the distribution of the galaxy at a specific large size.

The coming decade should let us shed more light on dark energy.

Chapter 6: Dark Gravity

Why is it called Dark Gravity?

Dark gravity refer to the limits of our understanding of gravity and space-time, more than the energy field of a particle like in the case of the dark-energy field and dark matter. Gravity is quite fragile in comparison to the other three factors, while gravitons have remained undiscovered to this day. Although Einstein's general theory of relativity is revolutionary in expanding knowledge of gravity and has proven to be extremely effective as a result of observations, it remains an old-fashioned theory. It seems to be applicable on large scales, yet does not deal with the smallest scales of physical phenomena or the highest energy levels. This is because there isn't a confirmed quantum theory for gravity though there are many efforts underway to develop one. It is anticipated that there is a to carry gravity's charge called the graviton. The graviton is predicted to be a zero-mass particle and spin 2 (the photon, too, is massless and is spin-locked at one). However, the graviton hasn't yet been detected.

Gravity is a lot, much less powerful than the other three forces in the natural world (the electromagnetic force as well as the strong nuclear force as well as the force of nuclear weak). This may seem counter-experiential, but it is true. The reason gravity is perceived as important is because earth's weight is vast. However, even a tiny magnet is able to overpower gravity and lift the object. The majority of matter is near to neutral, so electromagnetic forces are blocked. However, when you reach for something, you're fighting gravity through electromagnetic forces which are the source of chemicals that propels your hand , allowing your hand to grip the item.

Dark energy and dark matter seem to be real and can be found in the universe. The quantities of the two amounts are measured. On the other hand, darkness is an abstract notion that is a reflection of our insufficient understanding of the fundamental facts about gravity and space-time.

The Standard Model of Particle Physics

The Standard Model provides a framework that explains the particles we know and the three forces that are known with quantum descriptions , electromagnetism the strong force, as well as the weak force. There are various types of particles those that are the basic of different particles and force carriers are described in the illustration on the next page.

Figure 6.1. This is the Standard Model for elementary particles. Quarks reside in the upper portion of the chart, while leptons are located in the lower section and the force carrying particles are in the top column. The graviton isn't part in the Standard Model. (Credits: Fermilab, Office of Science, United States Department of Energy, PBS Nova, MissMJ)

All particles are characterized by certain quantities that are fixed like their mass at rest as well as their charges, spins and so on. Charges are expressed in units of proton and electron charge, which are -1 or +1, respectively. Although the electron is regarded as fundamental, neutrons and protons are believed to be composed of triplets of up quarks as well as down quarks, with charges of either +2/3 or -1/3[1010. The proton, for instance is made up of two up quarks as well as one down quark that is equivalent to the charges of +1. The neutron is comprised of one up quark as well as two down quarks. This results in a charge of 0.

In the same way as matter particles, there are many particles called force carriers. For electromagnetic force, the zero mass spin+1 photon is the primary carrier. In the weak force, there are three components: the W+ The W+, the W+, as well as the Z. These particles are massive in contrast to the force carrier particles that are used for the force of strong. There are eight particles acting in the capacity of force carrier for powerful force. They are called gluons.

Additionally, there is a force-carrier for gravity, also known as the graviton. It's thought to be void of mass and spin at a rate of 2. It has not been discovered and is not included in the Standard Model.

The Standard Model has been extended by a theory called supersymmetry. It proposes that there are additional particles that are companions to the well-known particles with greater mass. For instance an electron with a spin of 1/2 could have a companion that has spin 1 referred to as the chooseron. None of these supersymmetric particles has been detected however they could be observed with larger mass (energies) by an accelerator like the Large Hadron Collider particle accelerator. One of the most probable possible candidates to detect dark matter could be the most light supersymmetric particle. It is expected to be extremely stable and weakly interfering (interacting via gravity only along with the force of weak).

Although symmetry is often thought of as beautiful however, it is that breaking through symmetry which allows for the fascinating universe we live in today. If everything was symmetric throughout the day, then there could only one force and only one type of particle. Humans, planets, and stars surely wouldn't be here. Symmetry breaking is an fundamental physics principle and is the key to understanding inflating the phase after the Big Bang and many other crucial aspects of the development in the Universe.

Supergravity

Supergravity was first suggested in the year 1976 by Freedman, Nieuwenhuizen and

Ferrara. It's a fusion of supersymmetry (on localized scales) and gravity (general relativity) that is formulated in four dimensions of space-time. The possibility exists to develop supergravity theories in higher dimensions, too. There are many interesting connections with string theory, for instance the 11-dimensional supergravity, which is the most expansive dimension that could be only one type of graviton. A space that has 11 dimensions is considered to be a desirable feature of string theory too and has been proved by Edward Witten that 11 dimensions is the smallest number needed to encompass the unification of all three forces that are not gravitational. One of the major issues with supergravity is that it seems to require a large Cosmological constant (dark energy) which is in stark contrast to the evidence.

Amendments to General Relativity

Certain physicists, like Jacob Bekenstein in particular, have proposed changes to general relativity which is a tensor-based theory of four-dimensional space-time. The tensor's elements describe the manner in which the space-time metric is curvilinear due to the presence of energy and mass. Some proposed changes include adding vector and scalar field terms to the equations which could alter the behavior of gravity on very large scales. Other theories that alter general relativity, like theories called f(R) and f(R) predict the existence of a time-varying gravitational constant.

These theories of modified gravitation attempt to understand the cosmic phenomena which have led to the demands of dark matter as well as dark energy.

The observations of galaxy clustering measurements taken from The Sloan Digital Sky Survey provide very strict guidelines on the magnitude of these deviations from general relativity may be. There is no reason apparent currently to implement any of these adjustments. These changes do not tackle the issue of how gravity can be unified with quantum Physics.

Loop Quantum Gravity

Loop quantum gravity begins with the realization that time and space cannot be continuous after you reach a certain scale. Quantization of space happens at a distance of 10^{-33} cm, also known as Planck length. It corresponds to one billionth of a trillionth of one centimeter. It can be 20 order of magnitude less than the atomic nucleus. The time scale is around 10^{-43} seconds, which is equivalent to the period during the time that light crosses the Planck length.

There exists a comparable energy of this magnitude that is known as the Planck energy. It's around 10^{19} GeV that amounts to 10^{19} (10 billion billion) times higher than the mass energy of proton.

Within Loop quantum gravity (LQG) theories that space is quantized using the mathematical model called spin networks as explained in the work of Roger Penrose. This is a method of constructing space into pieces or tiles. Time and causality are incorporated by imposing sequencing rules on the manner in which tiles are laid out and interconnected. Loop quantum gravity can be defined using the usual three dimensions of space, and also in the time dimension. Ashtekar, Rovelli, and Smolin are the main contributors to and advocates of LQG.

If the LQG hypothesis is true then the universe might have had an Big Bounce from a compacted quantized state instead of an actual Big Bang from a singularity. LQG seems to be capable of producing inflation that is that is similar to that of the inflationary Big Bang model. It is also possible to demonstrate the positive cosmological constant acting as dark energy, which is derived from LQG.

LQG doesn't clearly define the unification of forces , such as supersymmetry and string theories. It is possible that some elements from both theories may be correct.

Strings and Branes

One major area of study is the field of string theory, which is a quest to unite all four natural forces including gravity, and give a quantum foundation to gravity. It proposes tiny one-dimensional string as the basis of the universe, and not particle particles. String theory focuses on the fundamental nature of reality and matter from the smallest into the Planck length. There isn't a single string theory, but it is a sub-field of physics, which includes a range of theories offering a vast variety of possibilities.

In the case of energies less than 1016 GeV String theorists claim that they are able to reproduce Grand Unified Theory (GUT) models that model the three quantum mechanical forces, that don't include gravity. It is important to remember that, like strings theory, there isn't a any one-stop, accepted GUT theory. There is a variety of different theories.

It is mathematically compatible when there are dimensions in space that are greater than three; it is self-contradictory in 10, eleven or 26 degrees dependent on the way it is constructed. One of the 10 - or 11, 11- or even 26 dimensions is time. Three are the familiar spatial dimensions. The rest of the 6 and 7 or even 22 could be obscure (dark) physical dimensions which aren't visible to us. It is usually due to the fact that other dimensions are compressed to be extremely small and much smaller than what we can see even with the best of technology[1111.

Different vibration patterns of strings that are one-dimensional appear in various particles. The theory is comprised of close and open strings (loops) as well as has been expanded to include membranes or sheets that have two or more dimensions. They are able to be rigid and flat as well as having vibration modes by themselves. The majority of open strings be secured to some or all membranes which we typically refer to by their shorter name of branes. The requirement to fix the ends of strings that are open to a brane is known as the brane's confinement (not meant to refer to sending them to a mental institution). Strings with closed loops, like gravitons, are able to exist free on a membrane. we refer to them as free of confinement.

While strings are generally thought of as microscopically small, the branes may be quite big, and could be to be as big as the whole universe in certain models. Since the strings of that universe are contained within the brane area, the group of forces or particles are bound to only move within the same brane. Some models contain several branes. The area between them is known as the bulk. Therefore, one may feel forces across the bulk, and also on a few branes. Other forces can only be felt in the specific brane.

A brane that has three spatial dimensions with an extent equal to equal to the dimensions of our universe, or even larger could have risen from a tiny patch in the inflationary epoch. Remember from chapter 3 that the universe is constantly growing in size by 100 times or more in all three dimensions throughout the short inflationary period. Or, as a possibility in the model that we'll discuss in Chapter 7, it might be two or more large-scale branes prior to that of the Big Bang, and they met and created our Big Bang event.

Models that incorporate gravity in additional dimensions

There are a variety of models that have been constructed which are based on developments in string theory however, they do not need an exhaustive string theory study. They include models created of Lisa Randall and others wherein branes are believed to be part of a larger bulk with higher dimensions.

One illustrative model has 2 branes. The first brane may include our four-dimensional universe in which the familiar matter as well as the three forces that are not gravitational are contained. Gravity could be concentrated on a second brane but be able to penetrate the entire bulk. In reality, it could reach all the way to our brane - our globe, but it would be much weaker (an exponentially inverse measure of the distance to which it is separated) than the brane that is its home.

Figure 6.2 Figure 6.2 - Our brane that is a gravity brane as well as the rest of. Our brane is a 4D world, is located on the left. A second brane, located on the right, is separate from ours by an additional dimension (fifth dimension). Gravity is extremely powerful at the brane on the right and its force is able to penetrate the bulk , but it weakens as distance increases in an exponential way. In our brane, it's far, far less powerful.

The force of gravity on the brane that is its home will be similar to the force of the other three forces that impact our brane. However, it would decrease in strength rapidly increasing exponentially as it moves away from the brane that it is home to. So the strength of our brane may be required 36 or more orders of magnitudes less.

Entropic and Emergent Gravity

Models of this kind are referred to as emergent gravity, the entropic gravity model or induced gravity. In these types of models gravity isn't an intrinsic force! Instead, it is an emergent force, or derived force, with the thermodynamic or statistical type because of the spatial granularity at very tiny Planck scales. In the past 40 years, Andrei Sakharov[12] proposed that gravity arises from quantum Physics in a manner similar as the development of hydrodynamics from molecular Physics. Gravity develops into a statistical feature of quantum systems. This is a direct consequence from the quantum scale space-time as explained within the LQG section previously mentioned.

Stephen Hawking and Jacob Bekenstein demonstrated that the black hole's horizons are defined by temperature and the measure of entropy (the measurement of disorder and information content within the system) as well as that the entropy ratio is proportional to the size in the area of the black holes. This idea could expand to include the observer-friendly universe in general.

In the beginning of 2010, Erik Verlinde published a paper that derives Newton's gravitational laws from the first principles of Entropy (the measurement of disorder within the structure) the theory of information and also holographic theories. Ted Jacobson in 1995 had earlier claimed that Einstein's equation for general relativity was the equation of states that can be derived by thermodynamic argument.

What's common to all these related concepts is the fact that gravity is a statistical thermodynamic consequence of the fundamental Quantum nature of time and space at the tiniest of scales.

Summary

Gravity that is weak or dark is crucial because it allows for a universe to have huge scale and lengthy life. If gravity was similar to the other three fundamental forces, then the first disturbance that created the universe wouldn't have prevented it by gravity from increasing to a massive scale. There wouldn't be stars, planets, and even carbon or oxygen. For illustration that gravity is at least as strong as the powerful force, 1038 times more powerful than what it actually is. Also, imagine a homogeneous, expanding universe, isotropically expanding with the same density like our current universe. In this scenario, the typical timescale of the universe could be 0.04 seconds, with the typical size would be only 13,000 kilometers, roughly equivalent to Earth's radius!

There's a lot we aren't yet able to comprehend concerning gravity, even despite the enormous achievement that general relativity has achieved. To unify the physical science between the three quantum forces as well as the classical gravity theory theories such as supersymmetry and quantum loop gravity, supergravity as well as string theory are currently being created. There is widespread support that a quantum particle exists referred to as the graviton, a gravity force carrier, with 2 spins and a mass of zero. It is believed that if the massless two-spin particle can be discovered that it is the graviton. Finding single gravitons using current technology is a nearly impossible venture, however.

String theory states that gravitons as well as gravity aren't bound to branes. This could be the reason why gravity is smaller than the three other forces, that are bound to the brane upon which our universe can be observed.

In emerging gravity theories gravity has become an essential force, but rather a consequence of the thermodynamic properties of space-time quantized. This is a different explanation for the reason gravity is weak.

At present, the focus of experiments is on the detection of new heavier particles using the LHC to test the supersymmetry of string theories and supersymmetry. Additionally, there are ongoing research for gravitational waves that are the coherent states of several gravitons. Gravity waves are produced around black holes and neutron stars. An example can be found in that of the LIGO gravity wave research which includes detectors in Washington state as well as Louisiana. Furthermore, LQG makes predictions about the time delay in gamma ray propagation coming from extremely distant sources because of the "quantum foam" or the irregularity of space at extremely tiny quantum scales. The first observations have been made, but remain still inconclusive as of today.

Chapter 7: What Exactly Is An Hologram?

If we use the laser beam and divide it into two. The first beam is directed towards the object that will hold the hologram, and then reflected off it. It will then go to the object that the image is recorded. A second beam is going to be directed to the area where the recording will be made as well as the area that the first beam was directed. The pattern of interference that beams create as they crossed one another is what is recorded.

Our vision is a kind of Hologram

If we take a look at the box, it's not translated into an actual box inside our brain. The neurons in our brains that are firing produce electrical waves. A pattern of interference is formed because of this. We are able to make an internal hologram of the object we are watching.

The 98% optic nerves in the cat's eyes can be eliminated and its intricate visual tasks are not impaired. However, the remaining 2 percent could still accomplish the task efficiently.

Our memories as an photogram

A study that was conducted on a lot of cats and mice involved the teaching of them to run through mazes. A sweet treat was offered to them after the course. The brains of the patients were removed through surgery. Different brain regions were removed. However, each time the animals were able arrive at their destination on time. This proved that regardless of the brain part was removed the brain was not affected by memory loss. It is believed that the brain operates as the hologram.

The people who have undergone brain surgery are not affected by any loss of memory. They are able to remember all their family members , as well as many other things. This implies the application of a type of holography in our memory.

Quantum computers derived from the holograms

Photons are a great way to make computers to solve complicated issues. Interferometers, however, can become useless if their components are in conflict and cease to function.

Holograms made by interferometers are able to stop their properties from changing, resulting in reliable computers. They can be utilized to perform quantum computations. Glass that is tempered can be used to lock these photons, thereby protecting them from environmental influences.

Do we exist in the Holographic universe?

The indigenous people believed that we live in an imaginary world. The same is true for certain religions in the world. Researchers are working in the present time about the universe being in fact a Hologram. Grids are believed to form our the hologram. Electromagnetic energy on the physical level is the process of forming our awareness. The hologram is believed to have a beginning and a conclusion too. As the grids collapse, the hologram ends. grids in use the hologram will end.

A test is being conducted within Illinois by physicists who work at Fermilab. The experiment is based on lasers and will determine if our three-dimensional world is really the result of a two-dimensional matrix.

Many scientists are today referring to the nature of things as an electronic computer. There are laws that regulate the functioning of all things. They are described in mathematical formulas. Nature performs its real-time computations to determine where the moon is. We can do our own digital calculations about where the moon might be.

Aspect, a physicist, Aspect discovered subatomic particles nature can converse with one another in not being capable of traveling more quickly than the speed of light.

If we're looking at the same fish that is in an aquarium however, we're viewing it via two cameras set at right angles, the pictures we are seeing will be two distinct ones. The fish is one , however the angles we see it from will differ. The movements it performs will be the same. Bohm gives this as an explanation for Aspect interpretation.

Bohm says that we can think of objects as distinct when they are connected with each other. We can see only a small portion of their real world. The two parts aren't separate however, they can be described as being part of a image. At the deepest level this reality might be able to create, everything is connected to the rest of the universe. Nature creates a web in which the electrons that reside in the brain of humans are linked to the salmon, stars that sparkle in the night sky, and every heartbeat on earth.

Superholograms can also allow us to one day go into the past and seek out the events that we have completely forgotten about. The right tools for operating the hologram are required for this to be attainable. The hologram can then become an enormous storehouse of everything described as beings of matter and possess energy. They could be snowflakes, quasars and blue whales, and gamma rays as well. Bohm is also adamant that other objects could be hidden within the hologram that we could be left to imagine in the moment.

Stansilov Grof was conducting research on the use of LSD to treat pshychotherapy instrument. The time came when he encountered an unidentified female patient who was believing she was reptiles from the prehistoric age. The patient also described her male companion with great accuracy, describing him as having an elongated, coloured patch over his face. Grof discovered later that the colored scales on the bodies of reptiles are a significant role in the stimulant for sexual arousal in females. The patient had no prior knowledge of reptiles.

Grof discovered that some patients shared a commonality with almost every animal species that is known to exist. Certain patients were able to provide descriptions of funerals in detail and Hindu mythology stories with which they had never had experiences. This particular branch of psychology was called transpersonal physical science. But if we consider it from a hologram angle this gives us an idea of the connections between each atom, organ and even regions that do not have any time-space divisions between them.

Lyall Watson observed a woman who, following an elaborate dance, could cause trees to appear and then disappear as well. Scientists may not be able to provide an explanation using an holographic model. This could simply be an illusion of some kind. We might not have created our minds with thoughts that can accomplish such feats.

If the universe was an hologram, it would have no limit in the ways we could alter the world. The power of our minds can manipulate spoons, and make objects move into the sky. Reality is an unfinished canvas on which we could etch whatever we choose to. It would give us the power of our right. Random events could be identified by the hologram, however.

After studying the properties of black holes and the limits to which matter and energy could contain information that scientists see our universe to be a two 3D surface that formulas are written on it. Our 3D view could be merely looking at the same objects from various angles like the fish mentioned above.

Principle of Holographic

String theories in conjunction with quantum gravity suggest that a specific area could be encoded on the border of that area. It's best to light up by codes that regulate what happens within that space.

The idea behind this principle was originally developed by the the thermodynamics of black holes. Volume is believed as an illusion.

It is therefore an hologram that has information glued to its boundaries.

Craig Hogan, a Fermilab scientist states that quantum fluctuations will be observed in spatial positions that will cause background noise to be detected through gravitational waves detectors. However, this has not been observed so far.

The holographic principle stipulates the existence of a limitation on the amount of data that is stored within a certain space, which is also an energy limit that is finite. It's similar to the black hole dynamics.

Can time move slower for moving objects?

The measurement of time is done by clocks. If something moving at that speed, then the speed of light will decrease. If two twins are working in the same field and one of them goes to space, while the other remains on the earth, they'll get older in different ways. When the space twin is born, at the age of 10, he'll have accumulated 10 years while the twin in earth will have gotten older at 32. The astronaut was traveling at speeds that were relativistic which led to his timer slowing. It is believed that time is affected by gravity force. The Earth's time tends to pass slower.

All kinds of clocks have gone through this study. The clock that is analog, the clockwork clocks biological clocks, electronic watches, and Atomic clocks as well. They all slowing down when we walk the path of a moving vehicle.

Black holes

Black holes are believed to be a specific area of space in which gravitational fields are extremely dense, encapsulating all matter and radiation inside itself, and blocking any escape. Even light can't escape. This is why black holes appear invisibly. Everything is squeezed to fit in a narrow space due to the high gravity.

The theory is that black holes be located in the center of galaxies. This is because they are absorb material from around them, as well as merging with similarly formed black holes. The border of this zone, where escape is difficult is called"the event horizon..

The existence of a black hole is evident from the way it interacts with other objects and electromagnetic radiation. Light is an instance. Our personal Milky Way galaxy has been estimated to have the black hole, which is approximately 4.3 millions solar masses.

The matter that is sucked towards these black hole will break apart and glow brightly in the process. These are the most brilliant objects within the Universe. Quasars are also known as quasars.

The gases that circulate around the black holes transform it into an electrical generator. These electrical jets are scattered out to billions of kilometers into space.

However, just like black holes exist , so do there are white holes as well. These are a type of hole that sprays matter and are lit up like fountains. The white hole can't be opened from the outside and is not as deep as it's black counterpart. Matter as well as light are able to escape it. The holes are also attracted by matter, just as any other mass. Whatever object falls into it is not able to be able to reach the white hole's event horizon. A white hole's event horizon from the past was believed to be transformed into an event horizon of a black hole in the near future. A object that is thrown into an open white hole will be able to attain its black hole horizon.

White holes and black holes are believed to be the same thing. They are interchangeable while reversing inside themselves by the thermal equilibrium.

Chapter 8: The Ether: A Story Of A Old

Theory

The concept of the ether originates from the principles of classical mechanics, where we see that light transmission through space requires an intermediary (ether) in order the light (light) is able to move through it or within it.

In the past, when trying to find an absolute system (stationary ether) various experiments, such as those by Trouton as well as Noble[1] conducted.

Additionally, a well-known research was conducted by Michelson along with Morley.

The idea was that in the event that the ether existed that an observer from Earth could be able to detect"ether winds "ether winds" and that its speed corresponds to the speed of orbit for the earth. In order to do this, a change in the pattern of interference should be observed using an optical interferometer like Michelson's when light is reflected by the device.

Sunlight splits into two beams using the mirror that is half-silvered and then reflections are reflected by two mirrors that are placed 90 degrees one opposite. Two light beams are returned through the mirror that is half-silvered to the telescope located at "T" which is where the pattern of interference has to change when the interferometer rotates 90 degrees.

The popular optical experiment was carried out by A. Michelson in 1881 while he was trying to determine the motion of the earth's orbit across space. There was no shift like this was detected.

The initial result was shocking and demoralizing. This experiment went on for a second time, and an even more precise attempt was made by D. C. Miller with the Michelson and Morley's instruments with greater optical trajectories. However, it produced a negative outcome.

Others hypotheses were put forward to defend being the existence of "the the ether" as well as for the purpose of explaining the negative outcomes of experiments using the optical interferometer. These include"contraction of body" proposed by "contraction of the body" that was proposed by Lorentz-Fitzgerald as well as"the "attraction of ether"[3[3].

This assumption suggested that the ether in contact with bodies of an undetermined mass, bonded to them, and therefore the fixed ether had no velocity when it came in contact with the bodies. In essence, each body carried its own local ether.

This was how the negative results of the experiment by Michelson and Morley was overturned. The idea for the attraction to ether also was dismissed when the well-known effects were considered; for instance, the stellar aberration as well as Fizeau's convection coefficient. [4]

There is no reason to doubt up until the present time to Einstein's unique theory of relativity. One of its main points is the invariance of the rate of light. This implies it is the case that in the absence of vacuum, the light speed is the same value as c across all inertial systems . It is also the reality that propagation of the light signal is not dependent on the source.

The previous paragraph is one of my worries that we all accept the constant speed of light, but as of now, there are any studies trying to clarify why electromagnetic waves, or light are at that speed, or what determines the speed of light.

On contrary we also know that the vacuum isn't empty. It is a swarm of dark matter as well as dark energy. So, a part of that dark energy might be the sought-after "ether" that allows or acts as a medium for transmitting light.

What kind of information is there in the

VACUUM?

Based on the invariance of velocity of light I've come up with a novel theory regarding how we can have a shared space.

To begin, I deny the long-held notions of years ago regarding what the essence of the ether is, such as those that claim the "the inertia" is an inactive system or made up of particles that are typically in a state of still (stationary ether) with zero mass that can eventually attract. Be aware that the properties that were attributed to the ether are bizarre: no density and flawless transparency[55.

The ultra-tiny particles, also known as "basic particles" that, it was believed, comprised the ether, would allow for communication of signals. However, this implies that the particles of atomic size that cause the disturbance or signal that needs to be transmitted generate that signal with the speed of propagation equivalent to c = 299,792,500m/sec. It was never believed that this was accurate because an infinity amount of energy could be required to send the signal to the infinity or "invent" particles with no mass to transmit the signals to the infinite.

As per the original concept, as "the ether" serves only as an intermediary, the signal has to be made or delivered into "the ether" in a manner like c. If otherwise, how would it be that these particles, standing at rest- could be able to transmit an information in the light speed?

In addition, not all signal is delivered to an object in the same speed as light. We are aware that particle atoms that interact with themselves or change their speed emit radiation (for example, electrons that move from one orbit to the next because of a deceleration , emits photons) and that the radiation travels at the speed of light, even if the atomic particles that produced the radiation were traveling at a slower speed than the speed of the speed of. Another instance is the electrons in cathode radiation that are moving at a slower speed than the speed of c which is reflected on a TV screen, but images that emit from the screen are with the same speed as the speed of.

So, we can conclude that in all of these various scenarios that an medium (the medium) experiences a disturbance and afterward the medium (the medium) propagates the disturbance at a rate of c. This is without having to prove that the atomic particles which created the signal were moving at speeds lower than the speed of light. This explains the fact that the medium has to have enough energy and speed- to take in an event and send it out at the rate of light.

I also have to say that I do not agree on this with Max Born, who emphasized the fact that "The nature of the elastic property was discovered more and more from electromagnetic forces and it is illogical trying the explanation of electromagnetic effects line with their elastic characteristics of any imaginary environment". [6]

My opinion of what actually happens in the world or medium or in the medium, or the ether (if you'd like) where the invisible matter is present is as follows:

The surrounding environment (the vacuum) is awash with what I refer to as "Basic Particles". We know that the vacuum isn't really empty since it's filled with Dark Matter, and also lots of Dark Energy.

Let me now discuss the issue of what these most fundamental energy particles could be.

The universe, which is the space around us is filled with a variety of types of "Basic Particles" smaller than the atomic particles we are familiar with.

The simplest of these tiny particles is that they move (they aren't stationary) and colliding with each with a constant movement like the molecules of gas. We can refer to this as"vibrating environments" "vibrating atmosphere".

The most important aspect that this model explains is the fact that these fundamental particles, the simplest ones - SBPare able to travel at themselves a speed. Furthermore, it is around 41% more than Light's speed, as we'll discover in the future. Furthermore, if we cause an amplitude in any area of the medium to create a wave and then if we follow a specific direction and determine the speed of the front wave created it would be found that the speed of the wave is equal to c.

The fact that the entire space that is not just the space within the particles, is filled with the fundamental particles and are moving at a rate of 1.41c across all directions (the vibrational surrounding) allows the following conclusion feasible:

A) The speed at which any body's relative speed to this environment can't be determined because the environment isn't in a state of equilibrium with respect (in relationship) with any other body or system within the Universe. The particles that comprise the vibrational environment are moving in all directions at simultaneously and the average velocity is 41% higher than the c.

The reason for this is that Michelson-Morley's test could not establish whether there is an intrinsic reference system The researchers, beginning with them, believed that it was in a state of equilibrium or was in a continuous motion in a certain direction. However, this environment is moving across all directions at the same all at once!

B) The particles can't be transported, nor identified as we will learn in the future, and

c) Any disturbance or signal at any point results in the SBP particles to alter their movement, causing them to transmit this change to their neighbors basic particles, and then, in turn, transmit this alteration to the particles that are close to them, transferring that perturbation to the next particle until it reaches infinity. with a velocity of c .

So, when this environment is notified of the signal (perturbation) then it propagates the signal at the speed that is characteristic to the surrounding environment, and also in all directions. therefore, the speed at which it propagates is always at a constant.

The rationale behind the invariance to the rate of light due to the dynamic environment that is created by moving particles. Mobile particles are continuously traveling and colliding into one with a speed greater than c.

The simple particles - SBPhave mass, that is constant. While these particles move at a faster speed than light but their mass hasn't increased until it is infinite and will never be zero at the point of rest (at the exact moment in when two of them collide in front-to-back).

Based on my theory, the mass variance is ascribed to all atomic particles, baryonic particles, also known as molecules which are bodies moving in the vibrational environment, but not for the fundamental particles that comprise the environment that vibrates. The fundamental particles could be colliding with one and each other for millions of years , and their velocity average will keep to be the same.

They are in absolute vacuum, and there is no friction that would stop their movements. They can be thousands of times more compact than electrons They are small, do not have internal components and are simply globes with the highest degree of strength in the universe.

I refer to them as "the "Simple Basic Particle" or "SBP".

There are many distinctions in radiation of electromagnetic wave in this type of environment, and the propagation of sound waves within gaz, there's some similarities. From a mechanical perspective standpoint, it is essential to recognize the connection with the velocity average of molecules in a gas (V) as well as the propagation of sound waves within the gas (S).

If we look at the table below (I created) we will be able to draw certain conclusions:

SOUND Velocity IN SOME GASSES

(At 20 degrees Celsius at the sea's at sea level (1 atm))

The GAS Average Velocity

(m/sec) Sound

Velocity S

(m/sec) Velocity

Factor (**)

V/S

Hydrogen 1,840.00 1,269.50 1.45

Helium 1,308.36 902.30 1.45

Water vapor (*) 615.38 410.00 1.50

Nitrogen 493.72 339.30 1.46

Air 486.00 331.45 1.47

Oxygen 461.60 317.20 1.46

CO_2 392.28 260.00 1.51

(*) The temperature of water vapor is regarded as 100 degrees Celsius.

(**) The variations in the velocity factor of one gas and another is because certain molecules are monoatomic while other are polyatomic. Therefore, the elasticity of their impact differs.

The sound's velocity is lower than the velocity of molecules because the molecules of gas that can transmit sound are moving in all directions, but not in the direction of reference that we want to determine how fast sound travels. It is extremely unlikely that all gas molecules which make up the front wave of sound travel in the same direction at any given time and that collisions between the following molecules are completely frontal and the list goes on. If that were the case in the future, the sound's velocity in hydrogen, for instance it would be about approximately 45% higher.

It is the case that the wave front is generated at a particular moment through millions of molecules which move every direction, and the average projection of all velocity vectors along a particular direction is exactly what is known as the sound speed.

When V is defined as the velocity of molecules in gas, the mean for the projected velocities within an arbitrary direction (taking into account only the molecules located on one perpendicular plane to the direction of reference) should be

V * Cos Ph * EF

The mean in The Ph Angle is about 45deg. The EF is the Elasticity Factor of collisions between gas molecules.

Thus, the sound speed S would be

The S value is equal to V 0.70711 * EF

Similar to electromagnetic radiation. The particles at any time make up what is known as the frontwave of electromagnetic signals are traveling across all directions (at speed of around 1,41c) and we have to project each fundamental particle's velocity vector within only one direction. Each particle's velocity vector has to be multiplied with Cos Ph. The elastic factor EF must be 1.

Many fundamental particles take part in the creation of an electromagnetic wave, each having specific directions at any given moment. A specific particle might be in the exact referent direction (Ph = 0.deg) at the moment of collision with the next particle. However, the second particle due to its direction, will bounce at an angle that is between zero and 90 degrees (with regards to the direction of the referent) in the event that, it is a result of the collision, it can contribute to an increase in the frequency of the waves. This means that it will rebound along the same side of an axis perpendicular in the direction that wave's propagation.

Our common sense suggests that we should think of the average of Ph (from zero from 0deg to 90deg) as the same as 45 degrees.

Thus, the difference between the basic particle average velocity VSBP as well as the electromagnetic wave velocity has to be

c = VSBP * Cos Ph

VSBP = c/Cos 45deg = c/0.70711 = 1.41c

So, the average speed of the particles that compose the vibrating environment should not exceed 1.41 c.

Imagine the amount of energy that the SBP is carrying, in a VSBP of 1.41 c. The small amount that the SBP has is one thousandth of an electron's mass, however the SBP is one of billions or billions of fundamental particles for every electron that exists in the Universe.

The most crucial thing is this: the medium has the capacity-with sufficient energy and speed to take a signal, that is transmitted at the speed of light.

The exact moment at the during the Big Bang this powerful environment most likely delivered a part of its energy to create the various particles of atomic matter we are familiar with. Then, after some time it was the time when the Universe was at equilibrium, and all particles of the atomic chain received their own energy- it is possible that it will fluctuate based on interactions between other atomic particles it is certain that at some point in the aftermath of the Big Bang all basic particles have reached an equilibrium state and they have maintained their energy since then and also, their current speed.

There are two more "Basic Particles" in the vacuum as we will examine in the next section.

MORE BASIC PARTICLES

To make complete the notion of dark energy and dark matter, we need to look at two additional particles that make up the vacuum. They could be an intermediate state between "SBP" and the other well-known elementary particles such as quarks and electrons.

The first I refer to as "BAP" is spherical and appears as if it is composed with three or two holes on the surface. The dimensions of these particles could be 20-30 times bigger than elementary particles. The holes on their surface seem to have been caused through other matter that struck them like large meteors that fell obliquely onto the earth's surface.

The result was like a cavity created by a tiny teaspoon that has removed the most material off one end than from the other and left an opening in the shape of tears.

The particles spin because they are surrounded by "SBP's" are beaten the cavities of each particle that are towards the identical direction. The energy received is converted into the rotating of "BAP".

At the final point, there's an equilibrium between the energy that is absorbed from SBP SBP as well as the energy that is generated by it by waves.

The waves that are produced by the chambers of these particles are similar to the waves that are produced by vacuum. When the BAP is rotating its cavities, they move forwards in a certain direction, and the elementary particles that entered each cavity will move in the direction of that. This leaves an empty space in their wake which is then filled by other fundamental particles, creating an emitted "vacuum" which propagates with the velocity of light. These particles create an appealing field, which expands to the limit of.

I call them they are the "Basic Attraction Particle" or "BAP" and they ought to be the primary ingredient in the formation of what's called dark mass. They could also be called "Gravitons" that are which are responsible for the formation of gravitational fields that atomic particles create.

Another kind of particle that is found in the vacuum space also has the shape of a circle; it is the same size as the BAP.

The particles are not round. They are like a thick piece with a twisted edge that form an S. They are particles which I will call "BRP" are moving rapidly, similar to an anemometer because the SBPs that collide with the concave portion of the BRP provide greater energy than SBP hitting the convex side of the BRP.

The particle, thanks to it spins, is able to achieve an equilibrium between the energy absorbed by it and that which it sends to its medium as the pressure wave produced from the two convex components that make up the particles. The pressure wave propagates with the velocity of light and in a radiating manner, eliminating all particles the point they are reaching creating a repulsive force which extends into the limit of.

This kind of particle is thought to be called a "Basic Repulsion Particle" hence its nickname "BRP".

Because of their size, they can't be identified, and due to the nature due to their nature and how they transfer energy and carry energy, they contribute to the growth in the Universe. Therefore, BRPs should therefore, constitute Dark Energy.

In reality, BRPs create a combined wave, and its principal feature is its repulsive nature , or "pressure wave" produced by the convex side of the BRP and then later, the weak vacuum wave created by its concave part.

In the end, I can think that there are three basic particle types.

The simplest and least significant of them all - the "SBP"will fill up all the space in the Universe that we consider to be empty. The energy of the SBP is far higher than that of all visible matter. The mass multiplied by the square of its velocity which is equivalent to 1.41c when multiplied by their numbers (billions of times more than all particles of visible matter) is an amazing figure. The particles don't communicate directly with the particle atoms we have, but interact with them by using"BRP" and the "BRP" as well as"BAP" "BAP".

It is likely that during the Big Bang these SBP probably were moving faster, possibly two times faster than light, and thus they were more energetic. However, they may have contributed a portion of that energy to fundamental attracted particles -"BAP"or dark matter, and also the fundamental repulsive particle -"BRP"or Dark Energy.

When all three particles,"BAP" "BAP" as well as the "BRP" attained the speed at which they rotate now, a balance was created between the energy of the three types of particle. The "SBP" has a speed of 1.41c and, thus is the definition of how fast light travels that we recognize.

"BAP" along with "BRP" particles will interact with larger particles, such as electrons and quarks. The particles -"BAP" as well as "BRP"send their respective signals to the "SBP" which is the source of all signals that they receive from neutrons, electrons and protons, regardless of whether they are magnetic or electrical.

In the sense that these signals are transmitted through "SBP" with the velocity of light the "BAP" or "BRP" which may be within the proximity of protons, electrons and neutrons. In turn, due to this, it is possible to transmit electromagnetic signals at the speed light becomes possible across the Universe.

I'm also of the opinion quarks and electrons must be able to have "BAP" within them, similar to gravitons that we are unable to detect, but which permit the particle to draw other particles in proportion to their size. This is in relation to the amount of "BAP" particles that could be contained within it. "BAP" particles are independent within particles of quarks and electrons.

"BRP" particles, on their own, might be able to enter the form of an electron or a quark butthey cannot remain inside due to its properties that force the particles to reject all things and at the end of the day they'll end up ejecting each other.

So it is possible that the "BAP" is the only of these fundamental particles that is independent of the other known particles like quarks or electrons and thus, impart the gravitational properties of these particles in addition to the mass properties that we are familiar with.

Let me think about the signals these particles generate around the particles. The waves of attraction or repulsion produced through "BAP" or "BRP" particles surrounding them don't seem to be continuous in nature, but more of an electronic nature. The reason for this is that other particles sense these forces in the form of pulses.

Therefore, we can consider that every "BAP" as well as "BRP" particle creates within them an appealing force field, or a force field that is repulsive. According to the figures, these force fields are not continuous in nature but are of a digital type. Each particle within those fields will experience a vibrating force that is either attractive or repellent as shown in the following images.

Due to the above you can be confident in saying that our world is either Quantic or Digital -if you wishbut it's not continuous.

Why can't basic information be identified or traced?

We've already stated that the fundamental particles that comprise this brand-new concept of world are of a size and mass that is lower than that of the most tiny particles of the atomic scale, for example the electron.

The fundamental particles are everywhere in spaces, including obviously, interatomic spaces. They transmit electronic signals that connect the nucleus with its electrons as well as between electrons in an atom as well as those of nearby atoms that are part of the molecule.

The fundamental particles that comprise this pulsating environment form the foundation of electromagnetic and gravitational fields.

Normally, SBP particles move around the same fixed area in space. Their journeys are extremely short , even though they are doing it at a greater velocity than light, since they rapidly find other particles of the same type that they can collide with and alter their path.

The continuous motion governed by chance in the area of atomic particles can be greatly altered, however, as the fundamental particles create the physical environment that supports electromagnetic interactions that permit the atomic particles to stay connected to the nucleus and electrons to stay in their orbits. The electric fields and magnetic fields result from the variations of the fundamental particle's motion and direction. The strength of these fields results in the reality of a higher density of the three basic particles (SBP BAP, SBP as well as BRP) at a particular point and at a given time.

Thus, the fundamental elements (the SBP) which comprise the earth's environment are typically located situated around or within the vicinity of a particular location in space. If a body like Earth is drawn near the earth, the various particles will pass through the atoms in the particles of the Earth and change their direction of movement in order to create an electromagnetic field for every particle that moves through it at each moment, changing their direction thousands and trillions of times staying in the same location after the Earth is gone. The Earth's environment isn't dragged.

The same thing happens across all "Basic Particles" in free space. In interstellar space, they are moving by chance, the movement is only altered by the radiation of light from stars, and also gravity fields that come from the stellar bodies.

It is not necessary for these elementary particles to change their current position regardless of the need to transfer the light of one galaxy the next. They just transmit their signal to the fundamental particles that follow and, afterward continue to collide with other fundamental particles. This is the way the electromagnetic or gravitational signal is able to displace itself into infinity without needing to be a "photon without mass" or the infusion of energy. Signals, once they have been in place for an moment, are independent of their origin.

It is also simple to see the reason why there is no atomic particle, and, consequently, the body or molecule cannot be moving at a faster speed than light.

This is because being atoms as well as molecules are based on the vibrational environment, whose constituent particles are C as their velocity limit in a particular direction. Additionally, it is important to note the fact that in order for an electron or particle of the atomic family to accelerate, electromagnetic fields must be applied to it. But, as the elements of this field are moving at a maximum speed which is equal to the value of c, whatever the intensity of the field is and its impact on the particle in question is not to cause its speed to a higher velocity than that of the field of.

If the atoms of the body are attempting to be moved at a rate that is faster than the speed of light, the fundamental particles will not have the capability of transmitting the electromagnetic fields inside every atom, nor the electromagnetic fields that exist between them, at least within the course of moving. Thus, the distance between molecules and atoms would decrease in this particular direction, as they approached that speed at which light travels and then they would break up when they reached the speed at which light travels.

It is also important to note that the equipment and elements used in laboratories to measure atomic particles work with ionizations or the traces created by these atomic particles due to collisions between other particles or their interactions to electromagnetic field. Atomic particles are characterized by an "field of matter" which is nothing more other than a certain configuration of the fundamental particles caused by the influence of the spin of an atomic particle, and the other properties of each particle's atomic structure like mass, precession as well as electric charges.

The "SBP" or the "BAP" or"BRP," or "BRP" particles can't be identified due to their dimensions. The instruments available are too big to detect these particles. They don't have any electric charge. They possess mass, but don't contain gravitons.

For instance, the "SBP's" are merely solid spheres made of dense material that are not impacted by any known field.

Chapter 9: Future Of The Universe

The Universe Today

Let's look at the current state of affairs. The universe is just less than fourteen billion years old. It's still in its infant stage when compared to the next generation.

The scale factor that determines the dimensions that the universe (in each of the three dimension of linear space) is simply a function of the redshift. We then normalize the scale factor by a, and then compare it to its current value, a0. This is the value that we have at the time of our current epoch when redshift Z = 0. It is important to remember that higher z means that we are reflecting backwards further into time. In other epochs, it is the ratio of scale:

a = a0 / (1 + z)

For instance when redshift 1 z which was around six billion years old its size was approximately half of the present size in all three spatial dimensions. Therefore, the size of our universe that we could observe was approximately 1/2 x 1/2 half the size of our current universe which is eight times smaller. The density of matter was around 8 times greater. At the time the universe's expansion was driven by dark matter. However, as the density continued to decrease the dark energy concept became dominant following the value of z = 0.4. This was approximately 5 billion years old.

To look at our future plans, consider the z value as negative. as it decreases to zero The size factor of the universe will increase until it is infinite.

It is important to remember that current measurements show that we only have 4 percent of the total mass-energy of the universe, in the form of ordinary matter, with 23% of it being dark matter and 73% of dark energy. The components are held together by the space-time fabric and gravity.

The Future

In the coming years it is expected that the mean matter (dark and normal) densities will decrease in relation to the expanding size of the space (proportional in proportion to the square of the scaling factor) however the density of dark energy remains constant for each area unit. This means that dark energy is dominant to an the extent that it does as the universe expands. It is predicted to rise from 73% of mass-energy to close to 100 to 100 percent. The hypothesis in this study is that L (the quantity of the constant cosmological also known as dark energy which is represented by the Greek capital letters lambda) remains constant over time, as evidence from observation suggests, however it can't be proven that it's changing or could change in the near future.

In the case of constantL the scale factor increases exponentially as time passes and the expansion of the universe increases dramatically, leading to an uncontrollable condition.

Be aware that galaxies do not change in dimensions or density, but the distance between galaxies in general is increasing rapidly. Galaxy clusters that are currently in a gravitational bound state and have fallen out of the overall Hubble flow, remain bound. However, the dominant role in the universe's cosmological constant, or dark energy term over the past few billion years has stopped the growth of new galaxies clusters.

Galaxies will join with their gravitationally bound counterparts over the course of billions of years. It is believed that the Milky Way and Andromeda galaxies could join within three billion years. At some point, whole clusters and groups will join to form a number of smaller but more massive galaxies. Our local galaxy group could coalesce before the 1 trillionth year has gone by. Because the expansion of the universe is speeding up dramatically because of dark energy, the creation of new clusters and groups is being slowed. The universe will be awash with massive isolated galaxies expanding and separating from their other non-gravitationally bound counterparts.

Our horizon, though increasing relative to its size, begins shrinking in terms of relative. The light cone's horizon determines the galaxies that are easily visible. As the expansion continues forward, and the exponential growth in the scale factor over time, more more of distant galaxies will be not visible to us regardless of the capabilities of our telescopes. In fact, the vast majority of galaxies that are distant are disappearing at a faster rate than light speed. In the present, only a tiny portion of the universe forecast by the inflationary theory of cosmology lies within our light cone's horizon. Keep in mind from Chapter 3 how relativity permits the space fabric to stretch out at more that the light speed, and this occurred to a massive extent in the initial inflation phase.

As the universe continues to accelerate its expansion, and after million years of time have gone by, fewer galaxies can be seen each one another due to the larger scale that the universe has. Our local supercluster of galaxies that comprises the known as the Local Group of galaxies, is currently around 100 million light-years across. After two trillion years, a person looking within the Milky Way would only be capable of detecting other members of the supercluster but nothing else than because of the extremely redshift. A society that is intelligent in the near future would only be able to observe one galaxy, their own because all the galaxies that make up their supercluster or cluster will have joined at an earlier point. With regard to their universe visible, that only one star would represent all that was left to observe.

Scenario A: Just one or two Trillion Years!

A trillion years equals 1000 billion years, which is approximately 70 times the present age of the universe.

Theoretical theories have been suggested that theories have been proposed that suggest the Big Bang is a result of the collision of two branes that are extremely large in three dimensions but only a small distance from each other in an additional spatial dimension. Our universe is located in one brane. Paul Steinhardt and Neil Turok believe that the collisions of these branes may occur in a cyclical pattern, which could result in repeated Big Bangs , possibly many trillions or more years between. A beautiful springlike force that acts as a dark energy could pull the branes apart initially, but very slowly and then , close to the collision, it would accelerate. After the collision, the branes' kinetic energy will be transformed into radiation and new matter. Energy is conserved over every cycle due to the changing gravitational energy to the energy of the branes' kinetics after they disintegrated due to every collision.

There could be tiny quantum ripples on the surface of the branes which means that the contact between branes could occur at different times at different places and result in areas of underdensity and overdensity. They would reflect within CMB. CMB and the dense regions will provide the seeds needed for the formation of galaxy and star stars.

Incredibly, this timescale of just a few trillion years is roughly in line with the time scale of the stars with the longest lives, namely red dwarfs. This is an example of perfect timing, a brand new universe might be born of a brand new Big Bang around the time all the stars in the universe of old were slowly dying out.

Scenario B A Googol of the years!

The number googol[14] was first formulated as a result of the mathematician's nephew Edward Kasner, and popularized by Kasner. It is also the number 10100. It is one trillion (1012) times 8 times, which is that is multiplied by 10000. A googol of years can be an extremely long time indeed. In writing, it's

10,000 x 1,000,000,000,000 1,000,000,000,000 one million,000,000 multiplied by 1,000,000,000,000 1,000,000,000,000 x 100,000,000,000 x 1,000,000,000,000 x 1,000,000,000,000,000,000 years.

In the runaway universe, the possibility of a googol-year period of time has to be considered. There could be various major phases, in accordance with Fred Adams and Greg Laughlin. In the googol-year timeframe they would be the primordial period and that of the star era (they refer to it as 'stelliferous'), the degenerate period, and finally, the black hole period.

The Primordial Era

We've already covered in chapter 2 as well as 3 the early period, which covers the beginning Big Bang and inflationary period that continues throughout the time of cosmic nucleosynthesis in the early moments of time in the universe. Also, it covers the beginning of cosmic microwave background radiation as the universe cools down to an equilibrium state around 380,000 years, and then ends with the beginning of star formation after thousands to hundreds of millions of years.

The possibility of life isn't present in the Primordial Era because there are no elements heavier than helium and a tiny amount of lithium are created in Big Bang nucleosynthesis. The way we live today requires nitrogen, carbon and oxygen at a minimum and hydrogen.

The Stellar Era

The stellar age covers the vast majority of the life span of the universe up to now and is expected to last far longer, even beyond 1 trillion years. This is due to the fact that low mass stars live for longer lives as compared to massive stars. Our Sun is an average "main series" star, however it is more large than the majority of stars.

In order to be considered a star, you must possess at least 8% or more of the mass Sun or more. This is the minimum requirement for a sufficient temperatures in the center of the star in order to start thermonuclear nuclear fusion, which transforms hydrogen nuclei to the helium nuclei[1515. The proton-proton cycle , as it's referred to transforms 0.7 percent of the total hydrogen atoms' mass in energies (E is mc2). The temperature required will be 10 million Kelvin or higher.

The life of stars is a cycle that includes birth and death. The rate of reaction for nuclear fusion is highly sensitive to temperature, and in an the form of exponential growth which is more intense as the temperature gets higher. Between 10 and 100 million K , the reactions that are relevant increase to several times the rate. The result is that this process overwhelms the more fuel that is available in bigger stars. The bigger the star is, the brighter it will appear and the shorter will be its life span.

Stars in"the "main series" are distinguished by a strong connection with their hue reflecting of their surface temperature as well as their brightness. Main sequence stars burn hydrogen into helium inside their cores. It is the length of this process that determines their lifespan. The brightness during the primary sequence stage is proportional to the mass increased to approximately 3.5 to 4.

Massive stars, that are blue when in an underlying sequence can have lives of just one million years. The total lifespan of the yellow Sun is approximately 10 billion years and we're about close to the end of that time since the Sun is just under five billion in age. A star that weighs one-tenth of the Sun is likely to be red, with a cool, sluggish surface. Its core temperature is less than the Sun's inner core. It produces hydrogen and Helium in a gradual and constant way, and has an estimated lifetime of one trillion years or more.

The final point of a star is determined by the mass of its stars. Each star sheds mass throughout their lifetimes however the most dramatic mass loss occurs with supernovae. They are the result of explosions of powerful stars, with masses that exceed 9x Sun's masses which happen at when they reach the end of lifespan when their fuel has run out. The remnant core could be either a neutron or a supernova or, if the initial star was larger than 20 solar masses then a black hole will be developed. Stars with less as 9 solar mass become the white dwarf as a remnant. They do not undergo a supernova However, they shed some of their mass to the interstellar medium particularly when they turn into red giants towards the final stages of their life span.

The stars with the most massive masses lose the bulk of their initial mass mostly during the last "death throes" of the supernova explosion and then recycle their materials in the interstellar medium. The material is able to form the form of clouds of cool gas that can be used for the creation and growth of stars.

Stars re-circulate a substantial part of their mass into the interstellar space. However, the creation of stars with low masses is more favored over high mass stars so that more and more mass is deposited in the more long-lived low mass stars. This is why the amount of new material to the interstellar medium diminishes over time, and the formation of new stars is occurring at a slower and more sluggish speed. It's been discovered that the duration to eliminate the majority of new star formation is around 10 trillion years, which is similar to the life span of the most long-lived and lowest mass stars.

The life cycle is thriving in the stellar era however only when that the next generation of stars begin to form. The first stars has only hydrogen and helium. But the huge stars of the current generation also burn some of their fuels through fusion reactions, resulting in carbon oxygen, nitrogen, as well as other heavy elements. The new elements are then recirculated in the interstellar medium, and then are integrated into the new generation of stars as well as their associated planets.

As time passes, the metal called heavy element quantity in our galaxy as well as other galaxies is growing. With the abundance of these heavy metals for exploration, planets with the appearance of Earth could be made. Planets with steady orbits about stars make ideal habitats for the development of life and thrive, especially if they are within a thermal zone, which allows the possibility of liquid water. It is now known that planets are frequent with over 500 being discovered outside our solar system. The rate of discovery of new planets is growing quickly.

The stellar era is expected to continue until all red stars of low mass are gone, maybe 10-100 trillion years which is around 11,000 times than the present date of. Although intelligent life requires billions of years to develop on Earth but the red stars of low mass with their long lives as high as 1000 times longer extremely stable environments for some of the planets that are part of their systems to evolve into intelligent life.

The next era will be dominated by the remains of burned out (dead) celestial bodies.

The Degenerate Era

It is at this time that the sky is going to turn black. After stellar formation is over and all remaining stars have ended their time being nuclear-fusion engine their remnants that are 'degenerate' remain. Around half of the total mass is located in white dwarfs. About 50% in brown dwarfs that are also known as failed stars and a small portion of the mass can be found in neutron stars as well as black holes. The picture below illustrates the star that has fallen off its outer layers, and whose core will eventually collapse and form white dwarf.

The term "degeneracy" in physics refers to not an inappropriate behavior, but rather to specific conditions of matter. For instance, if you have a brown or white dwarf, it's an electron degenerate material. For neutron stars the matter is called neutralon degenerate material. Degenerate matter is controlled by quantum mechanical rules instead of the more well-known classical laws for the pressure of gas and liquids.

Normal matter is characterized by its pressure, which is determined by its temperature and density. Degenerate matter's pressure is determined by the physical distance between the particles that make up its constituents such as neutrons, electrons or electrons that is, by density alone[16].

The Pauli exclusion principle in quantum mechanics is a strict rule that states that for particles referred to as fermions (those that have spins 1/2 , 3/2 or 3/2, etc.) No two particles are able to share an identical state. A state refers to the same quantum numbers as well as the same momentum and position to the extent of quantum uncertainty defined through the Heisenberg uncertainty principle. In layman's terms the two electrons or neutrons can't be in the same space. They must be spaced at a minimum. In the event that an electron in a specific moment has a certain speed and you decide to add an additional electron at that time and force it to change its momentum.

If the density is sufficiently high, the electron degeneracy pressure is dominant over thermal pressure and helps the star to resist gravity collapse. The density is enough that a white solar dwarf shrinks to approximately the size of earth. The maximal mass that can be supported by electron degeneracy is 1.4 times the sun's mass. Beyond that the remaining material will melt into the form of a neutron star.

Another kind of degenerate object includes the brown giant that is referred to as to be a protostar, (or failed star). They aren't massive enough to qualify as true stars, with masses ranging from around 1%- eight percent from the sun's mass. They are not able to combine hydrogen and the element helium within their cores due to the lack of core temperatures. However, they do possess some nuclear burning of hydrogen to the deuterium (heavy hydrogen). The energy released is a result of gravitational collapse in the main and they are of sufficient density to allow electron degenerate matter , as in the white dwarf. Brown dwarfs might be able to provide life to planets in their vicinity, however, they would have for it to have extremely slow chemical reactions.

Rarely, two brown dwarfs meet and create a new normal star, however since only a few exist in the degenerate period would be extremely faint.

In the case of neutron stars, at the process of tearing down the core during the supernova explosion, the majority of the protons and electrons are combined to form neutrons. massive quantities of powerful neutrinos release (this assists in driving the supernova explosion, which pushes away the outer layers of the stars). A remnant of a neutron-star is held in place by pressure from neutrons. Neutron stars are a lot smaller, having the mass that ranges from 1.5 -3 solar mass that is the area of a city. Beyond 3 solar masses, the pressure of neutron degeneracy is not enough to stand up to gravity, and a neutron star could collapse into an black hole.

Dark matter is not able to interact with dark matter however, over extremely lengthy time scales, about 10^{20} years dark matter mutual destruction will take place, particularly within the inner regions of brown dwarfs, where dark matter particles will be the most likely to be taken captive. White dwarfs will emit in the far infrared with very low power, around 1,000 trillion watts. The energy produced by this mechanism is not enough to maintain White dwarfs' temperature to lower than 100 Kelvins.

Due to the fact that there is an oversupply in matter over non-matter throughout this universe, scientists are aware that the baryon count is not completely maintained. This means that protons will degrade over a long period of time. We aren't sure how long the lifespan for the proton will be however, measurements taken from the Super-Kamiokande experiment (in Japan, as you might have guessed from the name) suggest it is at minimum 6×10^{33} years. Grand Unified Theories (GUTs) claim that the half-life is around 10^{36} years.

Protons within white dwarfs as well as brown dwarfs will decay in this time-scale and the stars would fade out. Neutrons within a nucleus composed of carbon or helium or other elements within the white dwarf or in those outer layer of neutron-star will also be decaying in a similar manner. Free neutrons decay over the duration of about 10 minutes, creating electrons and protons. Therefore, any neutrons that are free would swiftly change into protons and decay over these long durations.

In 1040 years, the entire nuclear nucleons (protons or neutrons) would have gone as would the entire universe be composed of mostly low energy radiation, neutrinos, electrons and Postrons. Additionally, there would be plenty of black holes, many millions per galaxy, and this brings us to the final time period we'll discuss that of the Black Hole era.

Black Hole Era

Typically black holes form when massive stars are at their end of life. They are the result of supernovae's imploding cores and are characterized by masses of 3 solar masses or more. A three Solar mass blackhole is about 9 km[17[17,18]. Black holes may also be produced in an binary star system where one of the stars is a neutron star , and it is able to accrete material from another part in the binary system. The falling matter may eventually cause the neutron star to exceed the limit and could become a black hole.

The second known kind of black hole can be located in the middle of our galaxy as well as other galaxies. It is generally only one is found per galaxy. A black hole could be anywhere from 1 million to the 10 billion solar mass. Black holes are the source of energy for galactic nuclei referred to as Quasars. A black hole can grow in size in the course of billions of years through taking in nearby stars within the galactic core. Because the radius of a blackhole can be measured in relation to mass. It will have the radius of a million to one billion or more times the size is the case for black stars. If a black hole with a solar mass of 50 million, it will have a radius that is equal to the Sun and Earth distance (1 Astronomical unit).

Another type that is a black hole could be known as the primary black hole. Primordial black holes could have formed in the Big Bang, and would be small objects that have a very low mass. There has never been any evidence of them.

Stephen Hawking showed that, contrary to the traditional perception of black holes always increasing in mass, they can also lose mass. In general relativity, there is no way to escape from a black space when it is close enough to what is referred to as an event horizon. However, general relativity, being a classic theory does not deal with the quantum nature black holes also have.

A quick thought experiment that invokes quantum mechanics demonstrates this is the reality. We can see from quantum mechanics that there exists a vast ocean of virtual particles that float swiftly into and out the universe "borrowing" energies of the universe. Virtual particles usually occur in pair. Imagine that an electron-positron pair has been formed exactly at the event boundary in the dark hole. One particle is headed towards the inside, while the other would be heading outward in order to preserve momentum. The particle that is headed outward in the event of having enough velocity, may escape and remain as a true, not virtual particle. The powerful gravitational shear or tidal forces near the event horizon can aid this process. Therefore, over time when these pair of particles were formed the black hole will release energy.

Hawking utilized quantum thermodynamics calculations to calculate the speed at which black holes emit, or shed mass. The radiation is comparable to that of the black body's thermal radiation however the greater the mass of the black hole the less the temperature. In the end, the time-to-live is proportional to volume that the black hole's mass. If the black hole is of stellar size the lifespan is extremely lengthy, and the effective temperature is very low, around one tenth of one millionth of one degree over absolute zero. A black hole could take 1067 years for it to completely evaporate! Three billion solar masses black hole that is, perhaps, the size of a massive black hole today located in the middle of galaxies, could require an astounding 1096 years for it to completely evaporate!

Primordial black holes however, would radiate their entire energy rapidly. A black hole with a mass of 1 Gram would go away by 10100x faster than an solar mass black hole and therefore would vanish within a tiny millisecond shortly following that Big Bang.

Black holes are currently absorbing much more power than they emit and emit, not just from falling matter as well as from the background of cosmic microwaves and stars that strike the dark hole. In the future due to the expanding of the universe and the expansion of space, these sources will not be able to provide much energy while the quantum-thermodynamic radiation coming from the black holes will prevail.

Because it is low, only extremely low energy particles can be emitted most of them very long wavelength photons as well as neutrinos with low energy.

A Way Out?

In Scenario A, we have an expanding or oscillating universe on the possibility of millions of years. However, the end result of Scenario B appears to be the so-called heat death (somewhat incorrectly named, since it is the result of a the absence in heat) in the Universe. In this scenario, all ordinary and dark matter will degrade into very small energy particle. All the light rays coming emanating from the background of cosmic microwaves and stars from old time and also from the black holes that are in decay are stretched to huge wavelengths. A large number of positrons as well as electrons are floating about. A large number of positron-electron pair would have been destroyed by gamma radiation, but these gamma radiations would also be stretched out in the course of time. All the electrons and positrons which remain would have created an atom of positrons and electrons (positronium) that were massive in size. Nothing is happening in 10-100 years (a googol-year) after the black holes have completely disappeared.

However, keep in mind that we have pretended that the cosmological constant was actually constant over a lengthy length of time. The cosmological constant, caused by the dark radiation of the vacuum could change to a different value.

We aren't sure what possible permitted values could be. We do know that it's tiny in terms of quantum particle physics while dominating the balance of mass and energy in the universe at present.

But the dark energy may have quantum tunnel[18] reaching a lower level. It's similar to radioactivity, where particles decay over a certain period of time dependent on what nature the particle concerned. It is known what an typical life span will be, however we can't determine when a particular particle would begin to decay.

This quantum tunneling procedure may result in a brand completely new physical set in a particular region of space. The area could expand into the previous universe. The release of energy due to the phase transition would result in the rapid growth of the phase of L. A new universe will emerge and undergo an inflationary process like we been discussing in Chapter 2. It could happen 1020 years, or 1040 years and 1080 years. There is no method of knowing.

It also provides a second possibility of a future and rebirth for this (a future) universe. In order to be the kind of universe we consider an interesting universe which could include life-forms, believe that dark matter dark energy, dark matter as well as dark (weak) gravity will each play a role.

Conclusion

Mystery of Dark Energy continues. The universe's expansion was first discovered in 1998. Dark energy is believed to exist, but hasn't been discovered directly. It is possible that it is not there in any way. But it does possess properties we haven't been able to discover in any other substance. This is the reason it's difficult to conceptualize and describe.

It is classified as an anti-gravity force. Dark energy is believed to have increased in size in the course of. There are however, still new theories like the one of Christos Tsagas that suggests our motion in the galaxy gives us an inaccurate perception of expansion. If we see something that we're moving it will appear as if it's moving too, however, when we move against it, we'll meet resistance.

Einsteins theories and equations are being put to the tests by scientists. They must be proven 100 100% right. Many studies are being conducted in this field too.

Time is viewed as going to grind to a stop by certain professors. It is possible that we are headed toward a revolving universe. The expanding universe could be a sign of the slow progress of time. Time began to emerge at the time that the Big explosion took place. It was just as it appeared but it could also disappear.

The mysterious substance known as dark energy continues to operate in peace. Scientists have determined after extensive experiments that there is an 99.996 percent possibility that it exists in the end. Even if we haven't been able to prove it down with a definitive conclusive theory, it's believed to exist.

Dark energy is believed to be the cause of areas that make up the background map of cosmic microwaves that seem to be more hot than other maps. The finding of Higgs boson, as well as other types of particles could also suggest that changes could be made to Einstein's theory of general relativity.

The next evolution of the cosmic microwave background will provide more information on this issue.

With all the advancements in technology, there is no way that the dark matter mystery can be solved in a hurry. This quest is one of the most thrilling and difficult one that scientists have undertaken.

Certain predictions were included in the texts of religions like the bible regarding the sun's dimming in the near future. Scientists are now in a position to explain these predictions using equations and measurements. Many think that the dark force might be God himself, since it is stated within the Bible that God is the force that is responsible for keeping the universe in its the right place.

This particle of dark matter by scientists is sure to cause an abundance of excitement as well as questions to theories. The search for answers to this mystery has led to many other discoveries, too.

The future is bright, with these exciting discoveries that will result in the development quantum computers that can help us make more advances in astrology and science.

Data gathered from all the sophisticated machinery created by hard-working scientists will provide more information about this mysterious dark matter in the near future. We will must wait and see the results.

Our discoveries could help us improve space travel. What is the next step that traveling through black holes might take us. It is possible that we can locate a quicker route to Pluto.

The required setup for investigation of dark matter demands a large amount of expert knowledge and also funding. It has managed to connect researchers from all over the globe to share their equipment. This will be extremely beneficial to the future generation of scientists since they'll have an already established design structure in collaboration to carry on their research.

www.ingramcontent.com/pod-product-compliance
Lightning Source LLC
Chambersburg PA
CBHW050402120526
44590CB00015B/1789